CYBER WARFARE ETHICS

Issues in Military Ethics

Edited by Don Carrick, King's College London, Michael Skerker, United States Naval Academy, and David Whetham, King's College London at the Joint Services Command and Staff College

With most officer training schools including military ethics as part of their programme, more than ever there is a need for clarity on ethical decision making. Contemporary military conflict is ever changing and with it military practitioners are confronted by new ethical challenges which often puts additional weight on the professional activity of personnel. At a minimum, military professionals need to have a clear knowledge of the laws that underpin their profession in order to evaluate situations quickly.

The series explores the complexities of acting ethically within the military system. It is not a philosophical debate on military ethics nor is it a general introduction. Instead, this series aims to provide real world guidance for military commanders and leaders. Edited alongside King's College London Centre for Military Ethics and the United States Naval Academy, this series brings a unique and relevant combination of practitioner and academic expertise which profoundly enhances the overall effect of the learning experience from its publications.

Other books in this series

Military Virtues
978 1 912440 00 9

Military Space Ethics
978 1 912440 29 0

Cyber Warfare Ethics

Edited by
MICHAEL SKERKER
United States Naval Academy, U.S.A
DAVID WHETHAM
King's College London, UK

Howgate Publishing Limited

Copyright © 2021 Michael Skerker and David Whetham

First published in 2021 by
Howgate Publishing Limited
Station House
50 North Street
Havant
Hampshire
PO9 1QU
Email: info@howgatepublishing.com
Web: www.howgatepublishing.com

All rights reserved.

No part of this publication may be reproduced, stored in a retrieval system, or transmitted in any form or by any means including photocopying, electronic, mechanical, recording or otherwise, without the prior permission of the rights holders, application for which must be made to the publisher.

British Library Cataloguing-in-Publication Data
A catalogue record for this book is available from the British Library

ISBN 978-1-912440-26-9 (pbk)
ISBN 978-1-912440-27-6 (ebk - PDF)
ISBN 978-1-912440-28-3 (ebk - ePUB)

Michael Skerker and David Whetham have asserted their right under the Copyright, Designs and Patents Act, 1988, to be identified as the editors of this work.

The views expressed in this book are those of the individual authors and do not necessarily reflect official policy or position.

CONTENTS

Notes on Contributors	*vii*
Foreword	*xiii*
List of Abbreviations	*xv*

Introduction   1
    Thomas W. Simpson

PART ONE

1   Just War Theory and Cyber Warfare   9
    Fritz Allhoff and Jonathan Milgrim

2   *Jus ad Vim:* Sub-Threshold Cyber Warfare   27
    Michael L. Gross

3   The Rights of Those Targeted in Military Cyber Operations   44
    Michael Skerker

PART TWO

4   Cyber Warfare and Conventional Military Operations   59
    Richard Schoonhoven

5   The Ethics of Cyber-Sabotage   74
    Jeremy Davis

6   Not War: The Ethics of "Phase Zero" Cyber Operations   92
    Edward Barrett

7	Ethics of Military Cyber Surveillance *Peter Lee*	110
8	Ethics and Cyber Enabled PSYOP *Adam Henschke*	129

PART THREE

9	The Morality of Machines: Cyber Guardian Angels *Andrew M. Tidmarsh*	151
10	AI Ethics *Scott Robbins*	172
11	Ethics and Cyber Systems: Artificial Intelligent Weapons Systems and Moral Slippage *Elke Schwarz*	189

Conclusion 205
David Whetham and George Lucas

Index *219*

NOTES ON CONTRIBUTORS

Professor Fritz Allhoff

Fritz Allhoff, J.D., Ph.D. is a Professor in the Department of Philosophy at Western Michigan University. He is a founding member of the Asia-Pacific Chapter of the International Society for Military Ethics, and the co-editor of Binary Bullets: The Ethics of Cyber warfare (2016). Dr Allhoff has been a Fulbright Specialist at the University of Iceland, focusing on cyber attacks and critical infrastructure; his work on cyber warfare has also been funded by the U.S. National Science Foundation.

Dr Edward Barrett

Edward Barrett is the Director of Research at the United States Naval Academy's Stockdale Center for Ethical Leadership. He completed a Ph.D. in political theory at the University of Chicago, is the author of Persons and Liberal Democracy: The Ethical and Political Thought of Karol Wojtyla/John Paul II and has written many articles/chapters on military ethics issues. Dr Barrett has served in the active duty and reserve Air Force as a pilot and strategic planner.

Dr Jeremy Davis

Jeremy Davis is currently a Postdoctoral Associate at the University of Florida. His current work, which is supported by the University of Florida Consortium on Trust in Media and Technology, focuses on the ethics of algorithmic systems, particularly in law enforcement, and the ways these systems affect public trust and institutional trustworthiness. He was previously a Visiting Assistant Professor at the United States Military Academy (West Point) and has taught ethics and philosophy at the Naval Postgraduate School and the University of Toronto. He received his PhD in

2019 from the University of Toronto, where he wrote a dissertation entitled, "National Partiality and War".

Professor Michael L. Gross
Michael L. Gross is Professor of Political Science at The University of Haifa, Israel, specializing in military ethics and military medical ethics and related questions of national security. His articles have appeared in the *New England Journal of Medicine, American Journal of Bioethics, The Journal of Military Ethics, The Hastings Center Report, The Journal of Medical Ethics, the Journal of Applied Philosophy, Social Forces, The Journal of Cyber Security* and elsewhere. His latest books include *Bioethics and Armed Conflict* (2006); *Moral Dilemmas of Modern War* (2010); *The Ethics of Insurgency* (2015), *Soft War: The Ethics of Unarmed Conflict*, edited with Tamar Meisels (2017) and the most recently, *Military Medical Ethics in Contemporary Armed Conflict: Mobilizing Medicine in the Pursuit of Just War* (2021) Michael Gross is editor of the Routledge book series *War, Conflict and Ethics* and has led workshops on battlefield ethics, medicine and national security for the Dutch Ministry of Defense, The US Army Medical Department, the Defence Medical Services (UK), The US Naval Academy, the International Committee of Military Medicine and the Medical Corps and National Security College of the Israel Defense Forces.

Dr Adam Henschke
Adam Henschke is an applied ethicist, working on areas that cross over between ethics, technology and security. He is a senior lecturer at the National Security College, at the Australian National University in Canberra, Australia. His research concerns ethical and philosophical analyses of information technology and its uses, military ethics and on relations between ethics and national security. Adam has published on surveillance, emerging military technologies and intelligence and cyberspace. His most recent book is *Ethics In An Age Of Surveillance: Personal Information And Virtual Identities* (2017). He is a commentator on issues of ethics and national security technologies in the Australian press and regularly gives executive and professional development classes to members of Australia's public service engaged in national security. Adam has been a senior researcher on the ERC advanced grant project Global Terrorism and Collective Moral Responsibility, a Brocher Foundation research fellow, a visiting researcher with the United Nation's Institute on Disarmament Research (UNIDIR) and a visiting professor at the University of Hong Kong. His work is currently

funded in part from an Australian Research Council grant, DP180103439: *Intelligence And National Security: Ethics Efficacy And Accountability*, and an Australia Department of Defence Strategic Policy grant, *Countering Foreign Interference And Cyber War Challenges*.

Professor Peter Lee
Peter Lee, Professor of Applied Ethics, is the Director, Security and Risk Research at the University of Portsmouth. His research spans the ethics of war, ethical and other human aspects of military, policing and wider security activities, and the ethics of autonomous weapon systems. In 2016 he was granted unprecedented research access to the two RAF Reaper (drone) squadrons for his book, *Reaper Force: The Inside Story of Britain's Drone Wars* (Paperback, August 2019). From 2008 to 2017 he taught ethics of war and law of armed conflict at Royal Air Force College Cranwell for Kings College London and the University of Portsmouth respectively. From 2001 to 2008 he served as a Royal Air Force chaplain.

Professor George Lucas
George Lucas is "Distinguished Chair in Ethics" *Emeritus* at the U.S. Naval Academy, and Professor *Emeritus* of Ethics and Public Policy at the Graduate School of Public Policy at the Naval Postgraduate School in Monterey, California. He has taught at Georgetown University, Notre Dame University, Emory University, Case-Western Reserve University, the French Military Academy (Saint-Cyr), and the Catholic University of Leuven in Belgium, and most recently served as the Vice Admiral James B. Stockdale Professor of Ethics at the U.S. Naval War College (Newport RI). His main areas of interest are applied moral philosophy and military ethics, and he has written on such topics as irregular and hybrid warfare, cyber conflict, military and professional ethics, and ethical challenges of emerging military technologies. Publications include *Ethics and Military Strategy: Moving Beyond Clausewitz* (2019), *Ethics and Cyber Warfare* (2017), *Military Ethics: What Everyone Needs to Know* (2016).

Jonathan Milgrim
Jonathan Milgrim is a Ph.D. student in the Department of Philosophy at the University of Washington. He has taught just war theory at several universities and has been active in the International Society for Military Ethics—including the North American, European, and Asia-Pacific Chapters.

Dr Scott Robbins
Scott is a Post-Doctoral Research Fellow at Bonn University in Germany. Scott recently completed his PhD in the ethics of artificial intelligence at the Technical University of Delft (title: Machine Learning & Counter-Terrorism: ethics, efficacy, and meaningful human control). He has a B.Sc. in Computer Science from California State University, Chico and an M.Sc. in Ethics of Technology from the University of Twente. He is a founding member of the Foundation for Responsible Robotics and a member of the 4TU Centre for Ethics and Technology. Scott is skeptical of AI as a grand solution to societal problems and argues that AI should be boring.

Professor Richard Schoonhoven
Richard Schoonhoven is an Associate Professor of Philosophy in the Department of English and Philosophy at the United States Military Academy at West Point. He has been teaching in the Department since receiving his PhD from the University of Michigan in 2000, where he trained as a philosopher of science. At the time of accepting the job at West Point, he knew nothing about military ethics; he knows only slightly more now, although he did serve as the Program Chair for the International Society for Military Ethics for six years, and currently sits on the board.

Dr Elke Schwarz
Elke Schwarz is Senior Lecturer in Political Theory at Queen Mary University London. Her research focuses on the intersection of ethics of war and ethics of technology with an emphasis on unmanned and autonomous / intelligent military technologies and their impact on the politics of contemporary warfare. She is the author of *'Death Machines: The Ethics of Violent Technologies'*, is an RSA Fellow and also a member of the International Committee for Robot Arms Control (ICRAC). Elke is also Co-Editor for the New Frontiers in International Relations Series.

Professor Thomas W. Simpson
Thomas Simpson is Associate Professor of Philosophy and Public Policy at the Blavatnik School of Government, University of Oxford, and a Senior Research Fellow at Wadham College. He works particularly on trust, and issues at the intersection of technology and security. He joined the School from Cambridge, where he was a Research Fellow at Sidney Sussex College, and was also educated (BA, MPhil, PhD). Between degrees he was an officer with the Royal Marines Commandos for 5 years. He served in

Northern Ireland; Baghdad, Iraq; and Helmand Province, Afghanistan. The academic life is undoubtedly a privilege, but he remains conflicted about its sedentary nature.

Professor Michael Skerker
Michael Skerker is a Professor in the Leadership, Ethics, and Law department at the U.S. Naval Academy. His research focuses on police, military, and intelligence ethics. Publications include articles and chapters on ethics and asymmetrical war, collective responsibility, police ethics, intelligence ethics, interrogation/torture, and the book *An Ethics of Interrogation* (University of Chicago Press, 2010). His most recent book *The Moral Status of Combatants: A New Theory of Just War* (Routledge, 2020) explores the moral status of combatants fighting in unjust wars. Prof. Skerker is also the co-editor of *Sovereignty and the New Executive Authority* (Oxford University Press, 2019) and *Military Virtues* (Howgate Publishing, 2019). Prof. Skerker is on the advisory board of the High Value Detainee Interrogation Group and consults with businesses, law enforcement, and the military.

Wing Commander Andrew M. Tidmarsh
Andrew Tidmarsh is a serving officer in the United Kingdom's Royal Air Force. A graduate of the UK Defence Academy's Advanced Command and Staff Course, a Cormorant Fellow and a Chief of the Air Staff's Fellow; Andrew has participated in operations in Europe, the Middle East and Africa. Having initially qualified as an Aerospace Battle Manager and later as a Typhoon Qualified Weapons Instructor, he has held positions in tactical units through to strategic headquarters. Andrew's post graduate experience includes achieving the degree of Master of Science in Aerosystems, Air Operations and Tactics; Master of Arts in Air Power in the Modern World; and Master of Research in Defence Studies. He is conducting his PhD research with King's College London's Defence Studies Department, exploring the potential for artificial intelligence to appreciate the ethics of war and to provide a moral anchor, or guardian angel, for its human participants.

Colonel Scott M. Virgil
Scott Virgil was commissioned as an Armor Officer in 1998 following graduation from the United States Military Academy at West Point. He has held various leadership roles throughout his 23-year career to include command of 1st Squadron, 33rd United States Cavalry Regiment "War

Rakkasans". COL Virgil's operational deployments include service in Afghanistan, Iraq, and Kosovo. He most recently served as the Director of the Simon Center for the Professional Military Ethic at the United States Military Academy at West Point, overseeing numerous character developmental programs to include stewardship of the Cadet Honor System, MX400 The Study of Officership, the Cadet Character Education Program, and the Honor, Trust, and Respect cadet committees. COL Virgil is now an Army War College Fellow studying at Columbia University.

Professor David Whetham
David Whetham is Professor of Ethics and the Military Profession in the Defence Studies Department of King's College London. He is the Director of the King's Centre for Military Ethics located at the UK's Joint Services Command and Staff College. David supports military ethics education in many different countries and has held Visiting Fellowships at the Stockdale Center for Ethical Leadership, US Naval Academy Annapolis, the Centre for Defence Leadership and Ethics at the Australian Defence College in Canberra and at the University of Glasgow. He was a Mid-Career Fellow at the British Academy in 2017-18 and is currently a Visiting Professorial Fellow at the University of New South Wales. He is a member of the UK MoD AI Ethics Advisory Panel, and in 2020 he was appointed as an Assistant Inspector-General to the Australian Defence Force to assist in the final stages of the Afghanistan Inquiry. David is the Vice President of the European Chapter of the International Society for Military Ethics (Euro ISME).

FOREWORD

It is a true honor and pleasure to write the forward for Cyber Warfare Ethics. When I think of ethics and war, my memory drifts across my 23 years of Army service and back to my earliest days as a cadet at the United States Military Academy at West Point studying the Law of Armed Conflict (LOAC). As with the many of my classmates, this period formed the foundation of, and indeed continues to shape, my profound understanding of the Profession of Arms. At the time, my law professors led discussions on international law that set conditions for war (*jus ad bellum*) and the conduct of war itself (*jus in bello*) including principles such as necessity, distinction, proportionality, and unnecessary suffering. As a cadet and young officer, I often analyzed these principles across rational state actors, a linear battlefield with few domains and clearly identified allies and enemies, and sometimes in a singular mindset of kinetic engagements.

At West Point we also studied Clausewitz's theories that the nature of war is timeless while the character of war changes over time. The immortal Clausewitz and his work from 200 years ago still rings true. The nature of war still involves a contest of wills played out across the human dimension and the political dimension, all within an environment of uncertainty. The character of war, however, continues to change, and we are now at a point where it is changing very rapidly in the cyber domain. This book brings together impressive contributions from a wide range of intellectuals and practitioners committed to the challenging debate concerning cyber advancements and warfare, placing the advancements in the cyber domain against the principles of the law of armed conflict while also acknowledging that the lines of combat, rational state actors, and clearly identified allies and enemies have all been blurred.

This book is a must-read for any leader that operates at the tactical, operational, or strategic level of war. In the not-too-distant future the cyber domain may very well become the main effort for implementation of national power with its ability to influence across the diplomatic, informational, military, and economic arenas. Young officers and non-commissioned officers will have access to a new stock of cyber and fully autonomous weapons, both kinetic and non-kinetic, and the ethical implications of using these weapons must be understood in order to provide proper purpose, motivation, and direction. At the strategic level leaders must address the future of the cyber domain within the context of *what can we do, what should we do, and what are other competitors doing* while also acknowledging that our competitors may view this topic through a different moral lens.

While the work of these impressive contributors will not resolve many of the burning questions facing the cyber domain and future warfare, the chapters attempt to establish an intellectual foundation to facilitate further discussion, debate, and refinement. We owe it to those that will fight our future conflicts to address these issues now and ultimately provide guidance to the warfighter, rather than push the burden of ethical responsibility upon the warfighter in the heat of battle.

Colonel Scott M. Virgil
Director of the Simon Centre for Professional Military Ethic, 2019-2021
United States Military Academy at West Point

LIST OF ABBREVIATIONS

ACD	active computer defenses
APT	advanced persistent threat
AI	Artificial intelligence
ATLAS	U.S. Army's Advanced Targeting and Lethality Automated System
ASPI	Australian Strategic Policy Institute
AWS	autonomous weapons systems
BUR	Bottom Up Review
CAD	computer-aided dispatch
CAT	Convention Against Torture
CYBERCOM	Cyber Command
DoD	Department of Defense
DDoS	distributed denial-of-service
DNI	Director of National Intelligence
GDP	Gross Domestic Product
GPS	Global Positioning System
HUMINT	human intelligence
IC	intelligence community
ICRC	International Committee of the Red Cross
ICT	information communication technologies
IHL	international humanitarian law
IP	intellectual property
IRA	Internet Research Agency
IS	Islamic State
ISR	intelligence, surveillance, and reconnaissance
LOAC	Law of Armed Conflict
MRC	major regional conflict

MOOTW	military operations other than war
ML	Machine Learning
MIRV	Multiple Independently Targetable Re-entry Vehicles
NERC	North American Electric Reliability Corporation
NSA	National Security Agency
NSS	national security strategy
OPM	U.S. Office of Personnel Management
PRC	People's Republic of China
PLC	programmable logic controllers
PSYOP	Psychological Operations
PTSD	Post-Traumatic Stress Disorder
PSAP	public safety answering points
RWA	right-wing authoritarianism
SDO	social-dominance orientation
SCADA	Supervisory Control and Data Acquisition
USB	Universal Serial Bus
UN	United Nations
UNCLOS	United Nations Convention on the Law of the Sea
WMD	weapons of mass destruction

INTRODUCTION

Thomas W. Simpson

The ability to conduct cyber operations is now essential for a military, and it is likely that, in future, it will become only more central to how soldiers conduct their business. But cyber operations raise difficult ethical questions. Even academic specialists are only grappling with these now, and for the serving soldier, the situation is more challenging still. Whether in senior leadership or responsible for delivering tactical effect, uniformed personnel will increasingly find themselves making consequential ethical decisions in terms of when and how they deploy cyber capabilities. This volume is an aid for those in this situation. Its distinctive contribution is in providing a guide to some of the knotty moral questions posed by cyber technologies, in a way which is easily-read and accessible to those in uniform—who want to make the right decisions, but who do not have the luxury of open-ended time to examine these issues, with their focus necessarily on operational effectiveness.

In this Introduction, I explore some of the considerations that make this is a timely issue, and provide a brief prospectus of the essays in the volume.

Why is Cyber Becoming More Central to Military Operations?

Over the last 15 years or so, cyber has moved from a niche capability which some militaries were experimenting with, to something much more central to its mission and effectiveness, which developed militaries cannot afford to be without. Various key points can be identified in this development, which brought the technologies involved to wider awareness. In Operation Orchard, in 2007, the Israeli airforce attacked a suspected nuclear reactor in Syria, with the Syrian radar system being fed a "false sky" picture. Also in 2007, distributed denial of service attacks from locations in Russia crippled Estonian online infrastructure, in retaliation for the relocation of a statue honouring Soviet dead from World War 2; in 2008, similar attacks were used to prepare for Russia's conventional invasion of Georgia. In 2009, the USA established CYBERCOM, a unified combatant command, bringing together military cyber capabilities from across Defence, and in 2010, Stuxnet was discovered, a piece of malware which had targeted the Iranian nuclear programme and was widely attributed to the USA and Israel.

Since then, digital technologies have only grown in importance, both for society generally and the military specifically. Digital technologies are now more pervasive, more powerful, and more relied upon than ever before. This will continue. The ability to use these technologies for one's own offensive purposes and, defensively, to deny an adversary the same, has become crucial. The need for militaries to have a serious cyber capability derives not simply from some features of digital technology, however. Deep trends in the co-evolution of war and technology also mandate such a capability. These deep trends are also reinforced by some more current, geo-strategic facts. Take these in turn.

Writing in 1899, Jean de Bloch anticipated the stalemate of World War One. 'There will be a belt a thousand paces wide, separating [two armies] as by neutral territory, swept by the fire of both sides, a belt which no living being can stand for a moment … [and] which no living being can cross.'[1] Bloch's analysis was prescient, being borne out by the war of attrition and minimal gains that was reproduced on the Western, Eastern, and Italian fronts. New technologies—most notably, artillery and machine-gun—had massively increased the firepower available. In conjunction with barbed wire, defensive forces now enjoyed a structural advantage over those attacking. Overcoming the defensive effectiveness of this firepower was the central challenge for leaders of the day. By the end of World War One, Stephen Biddle argues, "a process of convergent evolution under harsh wartime selection pressures had produced a stable and essentially transnational body of ideas on the methods needed to operate effectively in the face of radically lethal modern weapons."[2] Tactical experiments by German, British, and Russian commanders all contributed to this.[3]

The central innovation lay in bringing combined arms effects together, at a decisive place and time, enabling infantry to break through

---

[1] Jean de Bloch, *The Future of War in its Technical, Economic and Political Relations*, Boston: World Peace Foundation (1914), xxix-xxx.
[2] Stephen Biddle, *Military Power: Explaining Victory and Defeat in Modern Battle*, Princeton, NJ: Princeton University Press (2004), 28. I am grateful to Anusar Farooqui for this reference and the previous one.
[3] Mark Thompson, *The White War*, London: Faber and Faber (2008), 295.

the opposing lines, with low-level commanders then expected to take the initiative and exploit tactical success. Crucial to this was the ability to identify and exploit the enemy's weaknesses, and to do this repeatedly and more quickly than the enemy was able to react. This is the "OODA" loop; commanders Observe the situation, Orient themselves, Decide what to do, then Act. By doing so at speed, agile armies could overwhelm their more pedestrian opponents, disorientating their commanders, and leading to a collapse in the will and ability to fight of lower-level units. A seminal example of the success of this was the battle of Caporetto, in late 1917, in which the young Rommel was awarded the Blue Max for his decisive role in breaking the Italian army, on a front which had been nearly stationary for over two years. For conventional warfare, the central results of this process of convergent evolution a hundred years ago remain valid. Combined arms operations, aimed at an enemy's key point of weakness, with success exploited at speed, can bring about the collapse of an opposing military.

Cyber technologies matter, then, because they create both new opportunities and new vulnerabilities. In terms of opportunities, they create new ways to gather information, through sensors that can observe at distances not previously possible, or remotely, and through unobserved and hostile access to information, by hacking systems. With the growth in machine learning, modelling and simulation capacities, and more robust communication with greater bandwidth, able to relay more granular information, cyber technologies provide new tools for understanding what is going on and deciding how to act. And they provide new ways of acting: computer technologies make the firepower of platforms (tanks, aircraft, and so on) more precise, and perhaps in future also the infantry soldier's fire. Digital technologies do all this at a speed that is qualitatively distinct from previous technologies. Cyber thus creates a new dimension of effects, enhancing and supplementing those previously available, enabling a military to overwhelm its opponent. Each opportunity is also, however, a vulnerability. Increased reliance on networked computers creates new and more centralised points of failure, which enemies can exploit. Cyber technologies thus enable old principles of war to be implemented in new ways. All of this means that more uniformed personnel will, increasingly, find themselves fighting with a computer, rather than with a rifle, ship, or aircraft.

There are some more contingent reasons why militaries in developed countries are likely to continue to invest heavily in cyber, related to features of the geo-strategic order at the time of writing. The collapse of the U.S.-led ISAF mission in Afghanistan, in August and September 2021, has reinforced a wider perception that we are at a noteworthy hinge point in international relations. While much of the two decades since 9/11 have been dominated by a global counter-terrorism campaign, aimed at radical Islamism, that effort is receding in significance when compared with a new level of competition, and distrust, between the global powers of the U.S., the People's Republic of China, and Putin's Russia. An era of great power competition places two demands on a developed military. First, it must train and equip itself so that, in the worst-case scenario, it may fight in high-intensity, symmetric conflict. All tools that may give an advantage are therefore required, and cyber capabilities may prove to be the decisive difference between two adversaries. Second, because of the mutually-recognised downsides of armed conflict between such powers, ongoing operations are likely to take place using forms of force that fall short of war. Cyber is a paradigm tool for this, as it enables a military to act at a distance, non-lethally, and often with plausible deniability.

Ethical Challenges

For all these reasons, then, many serving military members are likely to find themselves using new tools, which are not those traditionally used by soldiers, sailors, or aircrew. Like all weapons, cyber capabilities raise difficult ethical challenges, surrounding when and how they may appropriately be used. As is well known, there is a long and developed body of thought which addresses the ethics of conventional force, collectively termed the Just War tradition. Being new, questions around the use of cyber technologies have not yet received the same degree of reflection.

These questions include the following. Are cyber-attacks equivalent to kinetic attacks, being a just cause for war (*casus belli*)? Who is liable to be targeted by a cyber-attack: should only combatants and their materiel be deliberately targeted, as with kinetic weapons, or is it permissible intentionally to harm non-combatants' interests through cyber-attacks? Who may be targeted by cyber influence operations, and what messages may be conveyed by them? What justifies the use of cyber-attacks outside of armed conflicts? Do cyber capabilities wrongfully lower the threshold for war, making war excessively likely? Given that computing technologies displace and may supplant human decision-making, who is responsible for their effects, and how should automated technologies be regulated and used?

These questions, and others, have in recent years been the subject of new and ongoing debate among academics. But this literature is largely inaccessible to serving military leaders, being both difficult to get hold of, and assuming that the reader is familiar with the detail of philosophical debates in ethics. The value of this volume is that it collects a series of essays that, both individually and collectively, relate the principles of Just War theory to novel cyber technologies, and in a way that is accessible to the non-specialist.

Guide to the Volume

The volume is divided into three parts. The first addresses foundational issues: what the principles of Just War theory are; what their basis is, and whether cyber-attacks may thus qualify as an armed attack, akin to a kinetic weapon (Allhoff and Milgrim); what this may imply for the use of force in contexts short of war (Gross); and what the limits on the use of force may be (Skerker).

The second part addresses a variety of more applied issues, looking at whether cyber-attacks are compliant with the principles of *jus ad bellum, in bello* and *post bellum*, or whether they pose new challenges (Schoonhoven); and then the ethics of cyber-attacks as forms of sabotage (David); espionage and shaping operations (Barrett; for the time-pressed reader, this may be a sensible place to start); surveillance (Lee) and psychological influence operations (Henschke).

The final part considers ethical questions raised by a specific application of computing technology, namely automation and machine learning. Perhaps machine learning systems may act benevolently, filtering out destructive influences on online communication (Tidmarsh). But, automated systems nonetheless pose ethical challenges, around algorithmic bias (Robbins), and ensuring that human operators are not just held responsible, but do not cede too much decision-making authority to machines (Schwarz).

The conclusion, by George Lucas, contains a fuller summary and discussion of the contribution of each of the papers.

PART ONE

1

JUST WAR THEORY AND CYBER WARFARE[1]

Fritz Allhoff[2] and Jonathan Milgrim[3]

Just War Theory

Just war theory comprises a long intellectual tradition, spanning ancient, Christian, medieval, scholastic, modern, and contemporary sources.[4] For our purposes, though, Michael Walzer's *Just and Unjust Wars* offers a good starting point.[5] One of the most central distinctions that we need to understand is the difference between *jus ad bellum* and *jus in bello*. As Walzer puts it:

> The moral reality of war is divided into two parts. War is always judged twice, first with reference to the reasons states have for fighting, secondly with reference to the means they adopt. The first kinds of judgment is adjectival in character: we say that a particular war is just or unjust. The second is adverbial: we say that the war is being fought

---

1 We thank Kirstin Howgate and Michael Skerker for helpful comments on earlier drafts. We also thank Edward Barrett, Shannon Ford, Adam Henschke, Patrick Lin, and George Lucas for prior conversations and collaborations.
2 Fritz Allhoff, J.D., Ph.D. is a Professor in the Department of Philosophy at Western Michigan University. He is a founding member of the Asia-Pacific Chapter of the International Society for Military Ethics, and the co-editor of *Binary Bullets: The Ethics of Cyber Warfare* (Oxford University Press, 2016). Dr. Allhoff has been a Fulbright Specialist at the University of Iceland, focusing on cyber attacks and critical infrastructure; his work on cyber warfare has also been funded by the U.S. National Science Foundation.
3 Jonathan Milgrim is a Ph.D. student in the Department of Philosophy at the University of Washington. He has taught just war theory at several universities and has been active in the International Society for Military Ethics—including the North American, European, and Asia-Pacific Chapters.
4 Gregory M. Reichberg, Henrik Syse, and Endre Begby (eds.), *The Ethics of War: Classical and Contemporary Readings* (Wiley Blackwell, 2006).
5 Michael Walzer, *Just and Unjust Wars*, 4th ed. (Basic Books, 1977).

justly or unjustly, Medieval writers made the difference a matter of propositions, distinguishing jus ad bellum, the justice of war, from jus in bello, justice in war.[6]

Putting it more colloquially, there are two different sorts of things just war theory asks us to think about. First, we have to determine *whether* we can fight. If so, we have a just war—note the adjectival usage. Second, though, we have to determine *how* we fight, for example, justly or unjustly—note the adverbial usage.

For Walzer, these are independent lines of inquiry. So, there are four possibilities:

1. We fight a just war justly
2. We fight a just war unjustly
3. We fight an unjust war justly
4. We fight an unjust war unjustly.

It is worth pausing here to note that we are basically one paragraph into just war theory—albeit a hugely important one—and we are already seeing a wide range of theoretical issues start to emerge. Importantly, Walzer thought that *jus ad bellum* and *jus in bello* could be assessed independently, a controversial notion owing to his views regarding the "moral equality of combatants."[7] So, for him, it is still possible for unjust aggressors to fight justly, because we separately evaluate the justice of the war itself (*ad bellum*) from the conduct within it (*in bello*). More recent theorists have disagreed and been more inclined to think that the unjust side inherently fights unjustly, thus denying Walzer's architecture.[8] But for our purposes, we shall treat *jus ad bellum* and *jus in bello* separately, while recognizing that this might be an oversimplification; regardless, it will allow us to begin to unpack some of the substantive issues that attach to each, as well as apply those issues to cyber warfare.

Most theorists agree that *jus ad bellum* has seven principal requirements: whether we are justified in turning to war depends on

---

6 Walzer, *Just and Unjust Wars*, 21.
7 Walzer, *Just and Unjust Wars*, 34.
8 See, for example, Jeff McMahan, *Killing in War* (Oxford University Press, 2009), especially § 2. More generally, so-called "revisionists" have put pressure on Walzer's "orthodoxy", though these can be technical debates beyond our current needs. For more discussion, see Seth Lazar, "War", *Stanford Encyclopedia of Philosophy* (2020). See also Helen Frowe, *The Ethics of War and Peace: An Introduction*, 2nd ed. (Routledge, 2016). Important starting points are: whether the soldiers know they are on the unjust side; whether, even if they do not know, they should know; whether they were conscripted or volunteered, and so on.

whether we can affirmatively satisfy all of these conditions. They comprise just cause; proportionality; reasonable chance of success; legitimate authority; right intention; last resort; and public declaration of war. Much indeed has been written about each of these, and debates abound about the proper interpretations of each. But the list itself is uncontroversial, as is the following general descriptions of the reasoning behind it:

- *Just cause* requires that we only go to war to defend against aggression.[9] We may not go to war to gain territory or wealth, but instead may only use force defensively. Humanitarian interventions may nevertheless be allowed if the intervening party is defending some other party against the aggression of a third party—generally subject to the sovereign will of the aggressed party.
- *Proportionality* requires that the costs of war not grossly exceed its benefits. Costs and benefits are broadly construed, including human, social, and economic.
- *Reasonable chance of success* requires that wars be winnable—it bars futile wars that would sacrifice resources (including combatants) for no good reason. If the war cannot be won, it should not be fought at all.
- *Legitimate authority* requires that war be waged by those whom are licensed to do so, generally heads of state. Wars may not be waged by private militias, sub-state actors, non-state actors, and so on.
- *Right intention* requires that wars are fought for the right reasons—specifically, to resist aggression. If some party would otherwise be justified in defending against aggression, but puts combatants in harm's way for some nefarious or pretextual reason, then the use of force is unjustified.[10]
- *Last resort* requires that reasonable alternatives be exhausted before turning to war—for example, diplomatic efforts, economic sanctions, UN Security Council resolutions, and so on. At the same time, last resort cannot be taken too literally because there would

---

9 Medieval authors listed more justifications, including punishment of wrongdoing, but the consensus in the last four centuries has focused on defense. We thank Michael Skerker for this point.

10 A standard example here Operation Desert Shield, where the U.S. intervened on behalf of Kuwait, against Iraqi aggression. Consistent with Kuwait's sovereign will, the U.S. would otherwise be licensed in third-party defense, but if the U.S. intervened to secure its own oil supply, rather than to defend the Kuwaitis against aggression, the right intention requirement would not be satisfied.

always be some option other than force—including waiting, doing nothing, and so on—and so the requirement here is one of good faith reasonableness.
- *Public declaration* finds support in the Hague Convention, which requires that war "must not commence without previous or explicit warning." Public declaration allows the citizenry to debate the war, and it also allows the war's possible target to pursue peaceful solutions.[11]
- By contrast, *jus in bello* generally comprises only two rules,[12] but they are both theoretically and pragmatically complicated—probably more so than the *jus ad bellum* requirements.
- *Discrimination* requires that force may only be used against legitimate targets. The quickest distinction is between combatants and noncombatants, but can be extended to structures as well.[13] Combatants are those who pose threats—whether direct or indirect—whereas noncombatants are those who do not. By extension, military structures may generally be targeted as well, whereas civilian structures may generally not be targeted. Prisoners of war must also be treated in accordance with the Third Geneva Convention, and may not be harmed.
- *Proportionality* requires that all tactics be militarily necessary, as well as dispense harm proportionate to the benefits they seek to secure. As an example, a nuclear weapon is likely to be both indiscriminate (per above) and unproportionate, because the damage is so massive; the U.S. attacks on Japan during WWII are appropriately controversial for both reasons—they may also not have been necessary, if a Japan surrender was imminent. Proportionality shows up as an *ad bellum* and *in bello* requirement, but of different application: the former regards the resort to war as a whole, whereas the latter regards the tactics within the war—those may sometimes separate.

---

11 Frowe, *The Ethics of War and Peace*, 66.
12 This taxonomy is common among just war theorists, but Frowe breaks it out into more detail, separating combatants, targets, tactics, and prisoners of war. Those map onto the principles I have described, but see Frowe (2016), 105 for discussion.
13 Note that most of the literature prefers the combatant/noncombatant distinction to the more colloquial military/civilian. What we ultimately care about—both ethically and legally—is who is liable to attack. And, as it turns out, there are both military noncombatants (for example, chaplains and medical personnel), as well as civilian combatants (for example, some private military contractors). So the combatant/noncombatant distinction is closer to capturing what needs to be captured here, though the military/civilian distinction will generally—if not always—track accordingly.

And so we now have at our disposal the key concepts and distinctions that will allow us to consider cyber warfare more directly. In the next section, we will unpack this framework, but mostly still on an abstract and theoretical level. In the third section, we will apply it to more concrete case studies.

Cyber Warfare

In this section, we shall consider how to think about just war theory in the cyber context.[14] We can gain purchase on this conversation by starting with a distinction between kinetic and cyber attacks, which will help get us situated. This distinction will also help quickly set up some of the differences between the two contexts, or at least challenge us to think through the ways in which we can extend more traditional thinking vis-à-vis kinetic attacks to the more emergent context of cyber warfare. Similarly, we can also begin to think through the ways in which these two contexts are analogous and disanalogous.

So just to get a simple case on the table, suppose that some state, A, launches a ballistic missile against some other state, B. Suppose that this missile was aimed at a crowded metropolitan center, and that various noncombatants are killed. Further suppose that there was an official declaration of war by A; in other words, we are not considering aggression by non-state actors. Here, it is uncontroversial that A has aggressed against B, and that B would be within its rights to exercise its right to self-defense, for example, against further attacks by A. B's entry into the conflict would still be governed by the other elements of *jus ad bellum*. A's aggressive act will underwrite B's invocation of just cause, but B must still tend to proportionality, reasonable chance of success, legitimate authority, right intention, last resort, and public declaration. Assuming that A can satisfy the rest of these, A will be justified in its resort to war.

But, focusing for now on just cause, how does this apply to cyber warfare? It is uncontroversial that a lethal threat from a ballistic missile comprises an aggressive act. But how do we understand "aggressive act" in the cyber context? Consider a range of possible cyber attacks:

---

14 This section draws from Patrick Lin, Fritz Allhoff, and Neil Rowe, "Is It Possible to Wage a Just Cyberwar?", *The Atlantic* (June 5, 2012). See also Fritz Allhoff, Nicholas G. Evans, and Adam Henschke, *The Routledge Handbook of Ethics and War: Just War Theory in the Twenty-First Century* (London: Routledge, 2013) and Fritz Allhoff, Adam Henschke, and Bradley Jay Strawser, *Binary Bullets: The Ethics of Cyber Warfare* (Oxford: Oxford University Press, 2016).

- A hacks B's government records, accessing millions of personnel files for government employees.[15]
- A targets B's economy, for example, by hacking banks, tanking the stock market, etc.[16]
- A targets B's internet infrastructure, for example, by distributed denials of service (DDoS's).[17]
- A hacks B's military infrastructure, for example, by either destroying weapons or knocking them offline.[18]
- A hacks B's critical infrastructure, for example, by targeting dams, electrical grids, aviation networks, etc.[19]

Which of these constitutes "aggression" under the terms of *jus ad bellum*? And what are the principles we invoke in order to figure it out?

More generally, are kinetic attacks and cyber attacks "commensurable?" In other words, we can imagine them being, broadly speaking, the same sorts of things—along whatever moral axes we are considering—or they could be different sorts of things. And if they are relevantly different, then how are we supposed to assess *jus ad bellum* in the cyber context? These sorts of questions can be illuminated by considering the oft-cited quote by a U.S. military official: "if you shut down our power grid, maybe we will put a missile down one of your smokestacks."[20] The sentiment behind this quote is one of *equivalence*, where the cyber attack is fully susceptible to a kinetic reply.

However, the extent to which that might be true probably depends substantially on the details of the cases that we are considering; in order to think through whether a kinetic reply would be justifiable, we would probably want to know a lot more about what this attack on the electrical grid looks like. For example, suppose that the cyber attack hacked the electrical grid, turning off power to hospitals—including life-saving ventilators. Alternatively, suppose the intervention against the electrical grid was more

---

15 See, for example, Devlin Barrett, "U.S. Suspects Hackers in China Breached about Four Million People's Records, Officials Say," *Wall Street Journal* (June 5, 2015).
16 See, for example, "NZ Takes Action over Stock Market Cyber Attacks," *BBC News* (August 28, 2020).
17 See, for example, Ian Traynor, "Russia Accused of Unleashing Cyberwar to Disable Estonia," *The Guardian* (May 17, 2007).
18 See, for example, Patrick Tucker, "Hacker Shows How to Break into Military Communications," *Defense One* (August 7, 2014).
19 See, for example, Mark Thompson, "Iranian Cyber Attack on New York Dam Shows Future of War," *Time* (March 24, 2016).
20 Siobhan Gorman and Julian E. Barnes, "Cyber Combat: Act of War," *Wall Street Journal* (May 31, 2011).

exploratory, like hacking into the control system of a small dam, but not actually opening the dam to flood the neighboring community.[21]

One intuition that we probably want to track from the outset is the difference between *lethal* and *sublethal* attacks. And so the equivalence between a cyber attack on an electrical grid and a kinetic attack on a smoke stack should probably trade on something other than bare sovereign or territorial disrespect. In other words, the mere fact that A targets the cyber infrastructure of B is probably inadequate to license resort to war, in much the same way that we would analyze sublethal kinetic threats (for example, throwing an apple at troops across a border). Conversely, if the cyber attack actually poses a lethal threat, there is no reason to treat the cyber context differently: being killed by a kinetic attack or a cyber attack is still being killed, and so we want to preserve the parity in those cases.[22]

While these have so far been philosophical musings, note that this broad approach is also captured under international law. As an example, consider Article 51 of the UN Charter, which circumscribes when member states may invoke their right to self-defense (for example, when the just cause prong of *jus ad bellum* is satisfied). It holds: "[n]othing in the present Charter shall impair the inherent right of individual or collective self-defense *if an armed attack occurs against a Member of the United Nations*, until the Security Council has taken measures necessary to maintain international peace and security" (emphasis added). In other words, while much of the document generally forbids the use of force—particularly before the Security Council would convene and issue a resolution—there is a substantial carve out for the allowance of self-defense in light of an "armed attack". This "armed attack" language gives us further purchase on what might qualify as an aggressive act under the *jus ad bellum* requirement of just cause. There are myriad interpretive issues, but, as a reasonable gloss, consider that "armed attack" comprises something like lethal threat.[23]

This then gives us a way to separate different sorts of cyber attacks, as well as a means by which to consider the equivalence with kinetic attacks. Note also, though, that there are still various other *ad bellum* requirements still at play. For example, suppose that we thought the attack on the electric

---

21 Thompson, "Iranian Cyber Attack."
22 See, for example, Melissa Eddy and Nicole Perloth, "Cyber Attack Suspected in German Woman's Death," *New York Times* (September 18, 2020).
23 For much more discussion, see Tom Ruys, *'Armed Attack' and Article 51 of the UN Charter* (Cambridge University Press, 2010).

grid was aggressive and provided just cause for the aggressed-against party. Still, whether that party's use of responsive force is governed by proportionality: putting a missile down a smokestack may be either proportionate or disproportionate, depending on the details. It is therefore important to reiterate that, just because just cause might be satisfied, it hardly licenses unlimited responsive force from the attacked party. And, as already mentioned, the remaining *ad bellum* requirements still need to be tended to as well.

But it is probably fair to say that, of the *ad bellum* requirements, just cause and proportionality are the ones most likely to pose interpretive issues for cyber warfare. Or at least insofar as it differs from kinetic warfare. The others—reasonable chance of success, legitimate authority, right intention, last resort and public declaration—are more straightforward. To be sure, there are all sorts of ways that those criteria can fail (for example, cyber attacks by sub-state actors will fail the criterion of right authority), but assessing the applicability of those criteria presents fewer novel issues than just cause and proportionality.

Turning to *jus in bello*, one of the principal concerns regarding cyber attacks is their tendency to be indiscriminate. For example, if hackers deploy a ransomware attack against a hospital, that would obviously be indiscriminate because it is not against a legitimate target. If someone goes on to actually die from such an attack, then that is even worse.[24] Many cyber attacks are going to look like this. From the above list, for example, almost all data, economic, and infrastructure attacks are going to substantially involve noncombatants.

Cyber attacks will tend to be more indiscriminate than kinetic attacks for a few different reasons. First, a kinetic attack will be physically delimited in certain ways: a missile, for example, can only do so much damage, and most of that damage is foreseeable given the missile's launch. So say state A fires a missile at some structure owned by state B. In this scenario, the structure might be destroyed, and we can generally wrap our heads around what that looks like. But instead, say that A interferes with an electrical grid for a city. There is all sorts of different stuff that could happen, much of which is barely foreseeable—though, at the same time, it is foreseeable that all sorts of bad things will happen, the point is just that we might not be able to predict, specifically, what they will be. Traffic lights being turned off, for example, makes road transportation substantially more dangerous.

---

24 See, for example, Eddy and Perloth, "Cyber Attack Suspected."

But it does not really necessitate any specific damage or injuries, unlike the missile that may have been aimed at a military target.

Second, the barrier to entry for a kinetic attack is significantly higher than it is for a cyber attack. If state A wants to launch a kinetic attack against state B, it needs weaponry, for example. Weaponry is expensive, takes up a bunch of storage space, can generally be detected from satellite imaging, and so on. Weaponry—especially of the sort that will be used in cross-state attacks—will generally accrue to state actors, who will generally be held accountable under both informal and formal sanctions (for example, everything from public outcry to international law, including Security Council resolution). We do not often have ballistic missiles appearing from nowhere, where we are unable to trace, for example, the launch site and launching parties.

Cyber warfare just looks different. They surely *can* originate through these sorts of "normal" avenues, but they can just as easily owe to a disaffected lone wolf hacker operating from his basement. There are obviously knowledge in terms of technical requirements, but if an 18-year old can hack the Pentagon,[25] that sort of shows where we are in terms of security flaws that might give hackers access to classified systems. Hackers might not have the same susceptibilities to sanctions that might deter state actors, which is another way of saying that they could be inadequately invested in practicing discriminatory tactics. Finally, the perpetrators of cyber attacks may not even be *known at all,* or else could be shrouded in the secrecy of groups of quasi-government groups (for example, as have been implicated in cyber attacks originating from China, North Korea, and Russia[26]) such that their identities are unknown altogether. That sort of anonymity compromises the accountability that can foster adhesion to *in bello* requirements.

But while cyber attacks may often be less discriminatory than kinetic attacks, we obviously should not rush to the conclusion that all are. There probably are "targeted" cyber attacks that are more discriminate, such as the U.S./Israeli collaboration against Iranian nuclear facilities— they released nefarious code that very specifically targeted the Iranian centrifuges, damaging them to the extent that the Iranian nuclear program

---

25 Anna Mulrine, "Meet David Dworken, the Teenager Who Hacked the Pentagon," *Christian Science Monitor* (July 5, 2016).
26 For more discussion, see George Lucas, *Ethics and Cyberwarfare* (Oxford University Press, 2016), 27-32.

was compromised.[27] There are also indiscriminate kinetic attacks, such as the allies' firebombing of Dresden during WWII or the U.S.'s nuclear bombing of Japan.[28] The point here is simply that cyber warfare is probably more indiscriminate overall, to the extent that it engages noncombatants more readily and more often than kinetic attacks.

As with all else, the *in bello* requirement of proportionality also raises various interpretive issues. For example, suppose that state A bombs state B. In this simple example, B could well respond with proportionate kinetic force against A—at this stage, we are in the *in bello* analysis, so let us assume the *ad bellum* requirements have already been satisfied. But suppose that, rather than respond kinetically against state B, state A reasons that B would be better off with a less belligerent leader, a sentiment broadly shared internationally. To what extent could A engage in election interference, for example, with the sole goal being to displace the existing leader? Setting aside other legal and international norms, we could easily suppose that this tactic would be *disproportionate* because subverting the entire democratic processes of an opposing state could be a lot worse, in terms of magnitude, than "merely" responding kinetically. The kinetic response, for example, would be localized in important ways, whereas the political destabilization could disenfranchise an entire electorate, its unfortunate political preferences notwithstanding.

The aim in this section has been to take the *jus ad bellum* and *jus in bello* requirements outlined above and contextualize them in terms of cyber attacks. In addition to showing how to think about them from the cyber context, it was useful to draw comparisons with kinetic attacks, showing the ways in which these analyses may converge and diverge. But as mentioned from the outset, this has still been a theoretical inquiry, broadly exploring various normative concepts. In the next section, we will turn to more specific examples, drawing richer, more empirically-informed examples, and scrutinizing them under the lens we have thus far ground.

---

27 Lucas, *Ethics and Cyberwarfare*, 58-60.
28 For more discussion, see Fritz Allhoff, *Terrorism, Ticking Time-Bombs, and Torture* (University of Chicago Press), 19-22.

Real-World Applications

In the previous section, we introduced several different kinds of cyber attacks. This section will provide real-world examples of each, and evaluate those examples through our just war theory framework. By considering these case studies, we will be able to further elucidate both the similarities and differences between cyber and kinetic attacks.

First, consider government A hacking government B's records, accessing millions of personnel files for government employees.[29] While there are several examples of this happening over the last couple decades, the most famous is the attack on U.S. Office of Personnel Management (OPM) by Chinese hackers. The attack took place over the course of almost a year, spanning the latter half of 2014 and the first few months of 2015. Attributed to hacker groups tied both directly and indirectly to the Chinese government, by the time the attack on the OPM was discovered and blocked the hackers had acquired a dizzying array of information. The information stolen by the hackers included 18 million copies of security clearance questionnaires, 4.2 million complete employee personnel files, and 5.6 million digital copies of fingerprints.[30]

While this sort of attack is concerning—especially for the millions of people who had their information stolen—it fails several *jus ad bellum* requirements for engaging in a just war. Most centrally, when considering proportionality, it is not clear how defensively attacking another country to stop the unjust acquisition of records can meet the requirement. The work of spies provide numerous non-cyber analogies for the theft of information. However, espionage is rarely, if ever, considered a just reason to engage in war.[31] While it is certainly unjust for a person's personal information to be stolen for whatever reason, it is difficult to imagine a scenario in which war is the appropriate level of response. The scenario would likely require that the hacking be centralized in an isolated location that is easy to target without causing great loss of life. Additionally, and more importantly, the scenario would have to involve the guarantee that using a kinetic weapon

---

29 Brendan Koerner, "Inside the Cyberattack that Shocked the U.S. Government," *Wired* (October 23, 2016).
30 For more discussion on the use of deception in cyberwarfare, see Heather Roff, "Cyber Perfidy, Ruse, and Deception," in Allhoff et al. (2016), 201-227.
31 For a fuller discussion of theft and espionage as it relates to just war theory, see Matthew Beard, "Just War, Cyberwar, and Cyber-Espionage," in Jai Galliot and Warren Reed (eds.), *Ethics and the Future of Spying* (New York: Routledge, 2016), 107-119.

against such a location would not lead to more widespread military activity. In the story above, we can likely imagine being able to discover a server farm or similar tech hub that was instrumental to the hackers success, and we can probably imagine being able to use very targeted force to destroy the location with minimal collateral damage. However, it is nearly impossible to imagine that this would not lead to much larger and more destructive actions between the Chinese and American militaries. It is much more likely that such activity will follow the path of spying and espionage, where all sides engage in the activity, all deny it, and it continues unabated mostly out of the public eye.[32]

A second form of attack involves government A targeting the economy of state B by hacking banks, attempting to disrupt the stock market, or similar means. Prior to one such attack in 2007, Estonia made the decision to remove a Soviet era statue honoring their efforts in World War II.[33] Russia warned the Estonian government that such disrespect would not be tolerated, but Estonia predictably ignored the warning. Shortly after the removal of the memorial to a local cemetery, Russian hackers launched a multi-pronged attack on Estonian banks and financial institutions, along with attacks on other areas of internet infrastructure. Estonia is one of the most internet-dependent populations in the world, and citizens soon discovered that the widespread DDoS (distributed denial-of-service) attacks limited or completely eliminated their ability to access nearly all parts of the internet, most notably critical banking functions. In a series of similar attacks from 2011 to 2013, Iranian hackers disrupted or attempted to disrupt as many as 46 financial institutions, including banking giants such as Chase, and credit card providers such as American Express.[34]

Again, this scenario seems to fail to meet *jus ad bellum* requirements for engaging in defensive war. While shutting down banks and other critical infrastructure could definitely reach a level of harm that appears to provide a just cause for defensive war—people starving due to economic collapse, etc.—and may satisfy the requirements for proportionality, it would require an extended period of time for such drastic effects to be realized. Long before these effects are reality, there is a host of options short

---

32 Lucas, *Ethics and Cyberwarfare*, 109-110.
33 Emily Tamkin, "10 Years After the Landmark Attack on Estonia, Is the World Better Prepared for Cyber Threats?", *Foreign Press* (April 27, 2017).
34 Dustin Volz and Jim Finkle, "U.S. Indicts Iranians for Hacking Dozens of Banks, New York Dam," *Reuters* (March 24, 2016).

of war that are available, from addressing the weaknesses of the affected networks to appealing to the international community.[35]

Another type of attack involves government A targeting the internet infrastructure of state B, through DDoS attacks and similar means. The attack on Estonia by Russian hackers was one such attack, and though the loss of banking ability was a major story at the time, the entirety of the Estonian internet infrastructure was affected. In February of 2020, the United States was suspected in a DDoS attack that crippled Iranian internet access for up to seven hours.[36] Countries that maintain strict control on internet content, such as Iran, are particularly prone to such attacks since there is generally less redundancy. At the height of the attack as much as 75 percent of Iranian internet infrastructure was affected. However, perhaps most interesting to our current discussion was the disruption of Georgian internet infrastructure during and in the days leading up to Russia's attack on Georgia in 2008.[37] At the time, Georgia was not heavily dependent on the internet. Certainly not as dependent as other countries such as Estonia and the United States. As a result, the attack mainly disrupted Georgian government websites, as well as transportation services. Georgian internet servers were originally attacked weeks before in what was described as a "dress rehearsal" for the coming war. Then, in the days before the Russian military moved against Georgia, the attacks began again. By the time actual bombs were falling, the Georgian internet infrastructure was already disabled. The key takeaway here is not the amount of disruption, but that it was a precursor to a kinetic attack by the Russian military.

If we consider these types of attacks in a similar context to our previous two examples, then they once again fail for a number of reasons such as proportionality and last resort. However, the conflict between Russia and Georgia provides another context in which to view these attacks. In the future, if such attacks are routinely used to disrupt communication, transportation, and similar services before a full-scale military attack, then they may rightly be viewed as an imminent threat.[38] It is not necessary to actually wait for the first shot to be fired, or for the first bomb to drop, in order to justly engage in defensive war—so long as the attack is imminent, it

---

35 For more discussion see George Lucas, "*Jus in Silico*: Moral Restrictions on the use of Cyberwarfare," in Allhoff et al. (2013), 367-375.
36 Alicia Hope, "Massive DDoS Attack Shuts Down Iran's Internet, Tehran Blames Washington," *CPO Magazine* (February 21, 2020).
37 John Markoff, "Before the Gunfire, Cyberattacks," *New York Times* (August 12, 2008).
38 For further discussion on imminent threats and preemptive attacks, see Henry Shue and David Rodin, *Preemption: Military Action and Moral Justification* (Oxford University Press, 2007).

need not (yet) be actual.[39] If such widescale attacks become the norm before an invasion or attack, then they may well be seen as the cyber equivalent of dropping paratroopers behind the enemy lines prior to the Normandy landings at D-Day. There might be no better way to gain an advantage in an increasingly connected world than to disrupt and disable a major part of their internet infrastructure. Cyber attacks could therefore be an opening salvo, a salvo that satisfies the previously problematic requirements to engage in a just defensive war.

The next type of cyber attack involves state A attacking the military infrastructure of state B. This may be by destroying such infrastructure, or, as the world becomes more dependent on equipment that is connected to networks, by knocking that infrastructure offline. We introduced such an attack earlier, namely the one on Iranian nuclear facilities by the United States and Israel.[40] This attack is different than normal cyber attacks in several ways. First, the attack was not a directed attack against computers, but a computer worm—called Stuxnet—that was designed to harm physical systems. Furthermore, the Stuxnet worm could not be placed in Iranian computers controlling nuclear equipment directly because the systems were not connected to the internet. Instead, the virus was placed on computers with which Iranian scientists would likely interact, such as companies in Germany that sold equipment needed by the Iranian nuclear program. Then, the virus was likely taken to the stand alone Iranian nuclear network via USB (Universal Serial Bus) flash drives by an unsuspecting Iranian scientist. The virus did minimal or no damage to most systems, simply existing on the drives until the correct system was detected. When placed on the correct system, it was designed to do a series of somewhat innocuous actions that, in themselves, did little immediate harm. It was also designed to report false information from the machines it infected. The result was that the Iranian nuclear centrifuges spun at a higher rate and valves were manipulated to increase the pressure inside the centrifuges. This caused the centrifuges to fail at a much higher rate, slowing down the Iranian efforts to enrich uranium. All the while the virus was reporting that the centrifuges were operating at optimal speeds and temperatures, leaving the Iranian scientists at a loss to explain the repeated failures.[41]

---

39 Allhoff (2012), 22-23, 167-68. See also Fritz Allhoff, "Self-Defense without Imminence," *American Criminal Law Review* 56.4 (2019): 1527-1552.
40 Kim Zetter, "An Unprecedented Look at Stuxnet, the World's First Digital Weapon," *Wired* (November 3, 2014).
41 Lucas, *Ethics and Cyberwarfare*, 58-60.

This attack represents something new and perhaps unique. It is hard to imagine a non-cyber scenario that is helpful in exploring *jus ad bellum* considerations. Iranian nuclear ambitions have been viewed with, to put it in the mildest terms, suspicion. However, even if we reimagine the scenario in which the Iranian efforts were for either civilian use, or for protection against an unjust aggressor, and imagine the U.S. as a potential aggressor acting in bad faith, this scenario still likely fails to satisfy the requirements for a just defensive war. First, there is the issue of imminence which undermines the last resort requirement. Then, there is the issue of proportionality, since the cyber attack on the nuclear centrifuges did no harm other than to slow Iranian advancement. Finally, there is the issue of a reasonable chance of success that lurks behind all these cases.[42]

The final example of cyber attacks involve state A attacking the critical infrastructure of state B, such as dams, electrical grids, or aviation networks. In the same attacks in 2011 to 2013 in which Iranian hackers targeted major banks and financial institutions in the U.S. with DDoS attacks, they also did something much less common and much more worrisome. Using a cellular network that was connected to a dam located in the state of New York, the Iranian hackers were able to infiltrate the software used to remotely control various parts of the dam, including floodgates.[43] By pure happenstance they were not actually able to control the functioning of the dam, but this was only because the gates themselves were manually disconnected from control hardware for maintenance. One worry with this sort of attack is that the systems themselves are particularly susceptible. There is no centralized command of such systems, and many are privately owned or controlled by local municipalities with little or no funding to protect against sophisticated attacks.[44]

This sort of cyber attack does have a sort of kinetic parallel. While the above example did not involve a dam actually being opened and allowing flooding, the potential is certainly there. During World War II, Operation Chastise involved British Royal Air Force bombers dropping "Dam Busters"

---

[42] For further discussion of the Stuxnet attack on Iran, see George Lucas, "*Jus in Silico*: Moral Restrictions on the use of Cyberwarfare" in Allhoff et al. (2013), 367-375 and Leonard Kahn, "Understanding Just Cause in Cyberwarfare" in Allhoff et al. (2013), 382-393.
[43] Thompson, "Iranian Cyber Attack."
[44] For a more detailed look at such threats, see Richard Clarke and Robert Knake, *Cyberwar: the Next Threat to National Security and What to Do about It* (HarperCollins, 2010) and Joel Brenner, *America the Vulnerable* (Penguin Books, 2013).

and breaching at least two dams in Germany.[45] As many as 1,600 civilians may have died as a result of the attacks, due to flooding of industrial centers and villages located downstream. It requires little creativity to imagine a similar scenario that involves not a bomb, but a compromised computer that controls floodgates. In such a scenario, it seems to matter very little if the damage is due to a kinetic attack such as a bomb, or due to a hacker successfully infiltrating critical infrastructure, the end results are the same.

Future Directions

After examining the various examples of past cyber maleficence, the important questions for our current discussion involve looking ahead to predict what these real-world examples might tell us about future conflict. There are two broad categories of cases that we have considered. Some cyber attacks pretty clearly do not meet either *jus ad bellum* or *jus in bello* requirements. Many of these have non-cyber analogies, such as spying and espionage. And yet others probably do satisfy the requirements. Attacks with destructive loss of life are easy to connect to standard kinetic attacks. A government hacking into and causing a failure at a dam, resulting in massive loss of life downstream, has a direct kinetic equivalent to dropping a bomb on the dam itself.[46] Similarly, from a standpoint of loss of life and justification for military response, there is little difference in hacking the guidance systems of aircraft and causing them to crash versus firing a missile with the same effect. However, between these two obvious categories is a large and expanding list of activities that fall into a grey area. Does hacking an election qualify? Destroying the economy of a region? What about the host of harms that are possible against individuals that, while constituting a harm, do not cause massive—or any—loss of life?[47] As we look to the future, these areas of discussion and these questions will become ever important as we seek to examine cyber attacks through just war theory.

---

[45] For further reading about this operation, see Max Arthur, *Dambusters: A Landmark Oral History* (Virgin Books, 2008) and Max Hastings, *Operation Chastise: The RAF's Most Brilliant Attack of World War II* (HarperCollins, 2020).

[46] See, for example, Dean Cheng, "What the Potential Crisis on the Yangtze Means for China and the World," *The Heritage Foundation* (August 5, 2020).

[47] One potential area of vulnerability that has not been discussed is massive number of household and other objects that are now connected to the internet. For more discussion on ethical issues, see Fritz Allhoff and Adam Henschke, "The Internet of Things: Foundational Ethical Issues," *Internet of Things* (2018) 55-66.

References

Allhoff, F. *Terrorism, Ticking Time-Bombs, and Torture*. University of Chicago Press, 2012.

Allhoff, F. "Self-Defense without Imminence," *American Criminal Law Review* 56.4 (2019): 1527-1552.

Allhoff, F. Nicholas G. Evans, and Adam Henschke, *The Routledge Handbook of Ethics and War: Just War Theory in the Twenty-First Century*. Routledge, 2013.

Allhoff, F. and A. Henschke, "The Internet of Things: Foundational Ethical Issues," *Internet of Things* (2018) 55-66.

Allhoff, F., A. Henschke, and B.J. Strawser, *Binary Bullets: The Ethics of Cyber Warfare*. Oxford University Press, 2016.

Arthur, M. *Dambusters: A Landmark Oral History*. London: Virgin Books, 2008.

Barrett, D. "U.S. Suspects Hackers in China Breached about Four Million People's Records, Officials Say," *Wall Street Journal* (June 5, 2015).

Beard, M. "Just War, Cyberwar, and Cyber-Espionage," in Jai Galliot and Warren Reed (eds.), *Ethics and the Future of Spying*. New York: Routledge, 2016, 107-19.

Brenner, J. *America the Vulnerable*. Penguin Books, 2013.

Cheng, D. "What the Potential Crisis on the Yangtze Means for China and the World," *The Heritage Foundation* (August 5, 2020).

Clarke, R. and R. Knake, *Cyberwar: the Next Threat to National Security and What to Do about It*. HarperCollins, 2010.

Eddy, M. and N. Perloth, "Cyber Attack Suspected in German Woman's Death," *New York Times* (September 18, 2020).

Frowe, H. *The Ethics of War and Peace: An Introduction*, 2nd ed. (Routledge, 2016).

Siobhan Gorman and Julian E. Barnes, "Cyber Combat: Act of War," *Wall Street Journal* (May 31, 2011).

Hastings, M. *Operation Chastise: The RAF's Most Brilliant Attack of World War II*. Harper Collins, 2020.

Hope, A. "Massive DDoS Attack Shuts Down Iran's Internet, Tehran Blames Washington," *CPO Magazine* (February 21, 2020).

Kahn, L. "Understanding Just Cause in Cyberwarfare" in Allhoff et al. (2013), 382-393.

Koerner, B. "Inside the Cyberattack that Shocked the U.S. Government," *Wired* (October 23, 2016).

Lazar, S. "War", *Stanford Encyclopedia of Philosophy* (2020).

Lin, P., F. Allhoff and N. Rowe, "Is It Possible to Wage a Just Cyberwar?", *The Atlantic* (June 5, 2012).

Lucas, G. "*Jus in Silico*: Moral Restrictions on the use of Cyberwarfare," in Allhoff et al. (2013), 367-375.

Lucas, G. *Ethics and Cyberwarfare*. Oxford University Press, 2016.

Markoff, J. "Before the Gunfire, Cyberattacks," *New York Times* (August 12, 2008).

McMahan, J. *Killing in War*. Oxford University Press, 2009.

Mulrine, A. "Meet David Dworken, the Teenager Who Hacked the Pentagon," *Christian Science Monitor* (July 5, 2016).

[no author], "NZ Takes Action over Stock Market Cyber Attacks," *BBC News* (August 28, 2020).

Reichberg, G.M., H. Syse, and E. Begby (eds.), *The Ethics of War: Classical and Contemporary Readings*. Wiley Blackwell, 2006.

Roff, H. "Cyber Perfidy, Ruse, and Deception," in Allhoff et al. (2016), 201-27.

Ruys, T. *'Armed Attack' and Article 51 of the UN Charter*. Cambridge University Press, 2010.

Shue, H. and D. Rodin, *Preemption: Military Action and Moral Justification*. Oxford University Press, 2007.

Tamkin, E. "10 Years After the Landmark Attack on Estonia, Is the World Better Prepared for Cyber Threats?", *Foreign Press* (April 27, 2017).

Thompson, M. "Iranian Cyber Attack on New York Dam Shows Future of War" *Time* (March 24, 2016).

Traynor, I. "Russia Accused of Unleashing Cyberwar to Disable Estonia," *The Guardian* (May 17, 2007).

Tucker, P. "Hacker Shows How to Break into Military Communications," *Defense One* (August 7, 2014).

Volz, V. and J. Finkle, "U.S. Indicts Iranians for Hacking Dozens of Banks, New York Dam," *Reuters* (March 24, 2016).

Walzer, M. *Just and Unjust Wars*, 4th ed. Basic Books, 1977.

Zetter, K. "An Unprecedented Look at Stuxnet, the World's First Digital Weapon," *Wired* (November 3, 2014).

2

JUS AD VIM

Sub-Threshold Cyber Warfare

Michael L. Gross

Jus ad vim is a current coinage that means "the right to use force." However, Michael Walzer introduced the term to refer to the right to use of force *short of war*.[1] The idea is to think about the kinds of force a state or non-state can exercise during peacetime without being drawn into a full-scale confrontation governed by the laws of armed conflict. One common example is a drone attack or border skirmish whereby states or non-states use armed force to gain limited political goals without breaching the threshold of "an armed attack." An example of the former might be Israel's limited missile attacks on Syria's T-4 airbase over the past several years. The ongoing dispute between India and Pakistan over Kashmir exemplifies the latter. In neither case are the countries currently at war, nor do the attacks necessarily draw the parties into conflict. These cases are interesting for Walzer because they represent the limited use of force about which there are no clear ethical or legal guidelines. They are interesting for us because they delimit the place of military-like force during peacetime. Cyber warfare is among the latest measures that fall into this category.

An "armed attack" is the threshold for war. Following such an attack, Article 51 of the UN charter affords nations the individual or collective right of self-defense. Simply put, nations may go to war to fight aggression. "Aggression" signifies rights-violating conduct. An aggressor violates the rights of a country or a people by significantly threatening the welfare of their political community. Invading deep into another nation's territory,

---

1 Michael Walzer, Just and Unjust Wars: A Moral Argument with Historical Illustrations. (New York: Basic Books (2006 [1977]), xv–xvi.

imposing a strangling blockade, or employing terrorism to deny a people self-determination and a dignified life are common examples. Absent successful attempts to rectify rights-violating behavior by other means, each violation permits recourse to armed force when necessary and proportionate.

The working definition of "armed attack" speaks to both the means and the consequences of an attack. The means are usually kinetic—bombs, tanks, missiles, and guns—while the consequences, writes Yoram Dinstein, must include "territorial intrusions, human casualties or considerable destruction of property."[2] It seems like the consequences are determinative. The border skirmish mentioned earlier utilizes kinetic weaponry but does not cause considerable destruction. On the other hand, a non-kinetic attack, and here cyber warfare is a prime example, can cause extensive devastation. Nevertheless, it is often hard to consider a cyber attack as an armed attack, particularly since cyber attacks have yet to harm anyone physically.[3] Beyond that, cyber measures do not obviously fit the definition of "armed" when cyber attackers steal information, destroy data, disrupt infrastructures, or manipulate the media.

So the questions before us are these. First, do cyber attacks constitute an armed attack that triggers the Geneva conventions, international humanitarian law, and the law of armed conflict? The short answer is no, not necessarily. Although cyber operations may rise to the level of an armed attack in the future, they need not. As such, they are a legitimate peacetime tool of politics as long as they remain below the armed-attack threshold. I will call this "sub-threshold" cyber warfare. And while cyber "operations" might be the more appropriate term, I deliberately use "warfare" to emphasize the coercive dimension of cyber campaigns. With that, the chapter turns to questions about the goals, means, and appropriate responses to sub-threshold cyber warfare. Among the ends and means, we can think of cyber attacks on critical infrastructures and cyber-mediated manipulation of the media or financial institutions to disrupt the flow of

---

2 Yoram Dinstein, War, Aggression and Self Defense. 4th edn. (Cambridge: Cambridge University Press, 2005).
3 On 18 September 2020, German police opened a murder investigation after hackers disrupted hospital service in a Ransomware attack in Düsseldorf. After hospital systems failed, one woman died as she was transferred to another hospital for life saving treatment. "Former chief executive of the UK's National Cyber Security Centre Ciaran Martin said: 'If confirmed, this tragedy would be the first known case of a death directly linked to a cyber-attack.'" Joe Tidy, Police launch homicide inquiry after German hospital hack, BBC News, https://www.bbc.com/news/technology-54204356.

information or money to hostile parties. Appropriate responses by states or non-states (such as guerrilla groups or liberation movements) may be preventive and include surveillance, for example, or retaliatory and include cyber or kinetic counter attacks.

As we consider these questions, we must also assess necessity, effectiveness, and the rights of attackers and targets. I use the last two terms neutrally and without attributing culpability. Attackers are those who initiate a cyber attack, while the target signifies those on the receiving end. Necessity asks, "Are there alternative measures to attain the same goal at lower costs?" If so, effectiveness then asks: is there a reasonable chance that cyber warfare will succeed? Nevertheless, the identity of the attackers and targets ultimately compels us to ask about culpability and liability. Do some attackers have stronger rights than others to wage cyber warfare? And, are some attackers or targets more culpable and less innocent than others? By addressing culpability, we realize that not all attackers are aggressors and not all targets are defenders. Sometimes attackers are not culpable while at other times defenders' actions may not be justified. At the same time, we may also ask how the inability to attribute responsibility or readily identify attackers or targets impinges on our assessment of the ethics of cyber warfare.

Attribution for cyber attacks is essential to assess proportionality. When unknown agents interfere with elections, disrupt air traffic control, steal proprietary information, or destroy data, we are far afield of the law of armed conflict. Without knowing the attacker's identity, how can we know if the target responds justly or proportionately? Proportionality weighs military advantage against civilian deaths, but there are no civilian deaths in these cases. Distinction demands attention to civilian immunity, but immunity only extends to injury, loss of life, and extensive property damage. Cyber attacks may cause none of these but still be violently disruptive. In this environment, we will see that proportionality means like-kind tit-for-tat responses rather than the overwhelming response typical of armed conflict. As such, sub-threshold cyber warfare is symmetrical even when the sides are not, and can level the playing field in a way conventional warfare cannot. These questions dog us throughout.

Do Cyber Attacks Cross the Threshold of an Armed Attack?

The cyber threshold problem is the opposite of what we usually expect. Think about drones. A drone missile attack clearly seems like an armed attack, and we very often have to explain why it does *not* pass the threshold of an armed attack and should not necessarily precipitate war. The same thing goes for low-intensity missile attacks that Hamas launches against Israel. Taken individually, these missiles are not sufficient cause for full-fledged war, but may warrant a limited armed (or cyber) response. However, as missile attacks accumulate, they may cross the threshold of an armed attack and invite powerful retaliatory strikes.

Cyber operations present quite the opposite problem. Cyber operations do not seem to approach the threshold of armed force at all. To date, cyber operations have not caused any deaths or injuries, although many scenarios depict mass casualties. One can imagine an attack on a hospital system or a critical water or transportation infrastructure that brings many casualties. But that has not happened yet. Suppose by warfare we mean operations that cross a threshold of devastation and loss of life accompanied by some organized clash of regular and irregular forces. In that case, the term "cyber warfare" is inexact. Cyber warfare does not seem to cross the necessary threshold of devastation or bear the marks of organized confrontation. Directed at civilian targets by agents unknown, cyber warfare is a kind of non-lethal terrorism that threatens to disrupt civil society by stealing information, manipulating the media, unsettling financial markets, or upending the electoral process. These operations hardly sound like the exercise of force at all.

The Psychological Threshold of Armed Attacks and War

Failing to cause significant physical destruction or injury, however, is not the only metric of an armed attack. Rightly so, Tallinn Manuals 1.0 and 2.0 on the *International Law Applicable to Cyber Operations* understand that some cyber operations may cause "severe mental suffering" that is tantamount to injury. In particular, they note the mental suffering that attends terrorism. As such, any cyber attack that causes severe mental suffering is on par with an armed attack. Anything less is a sub-threshold attack.

Exact definitions of *severe suffering*, mental or otherwise, enjoy no consensus and are absent from international law. International law, for example, bans weapons that cause superfluous injury and unnecessary suffering. The Convention Against Torture (CAT) prohibits acts causing "severe pain and suffering" but never defines any of these limits with precision. Instead of providing criteria for suffering, the law is content with banning certain weapons that seem inhumane (for example, blinding lasers or poison gas). And while some have sought to quantify torture in terms of the duration and after-effects of different interrogation techniques, CAT focuses its prohibition on the purpose of torture, namely to obtain information or confessions, rather than its specific means.

So it is with severe mental suffering. One idea is to work backward from terrorism. Thinking we know terrorism when we see it, we can then ask how it affects people. Here the literature paints a singular picture of anxiety, depression, dysfunction, fear, anger, xenophobia, recurring memories of trauma and, in the most acute cases, post-traumatic stress disorder (PTSD). In all or part, these outcomes offer a succinct picture of severe psychological harm. Does cyber warfare cross this line?

Cyber Warfare and Economic Warfare: Sub-Threshold Cousins

As blurry as it is, it is essential to keep the just-described threshold of armed-attack in mind as we assess the limits of cyber warfare. It is particularly germane when I compare cyber operations to economic sanctions. Generally speaking, economic sanctions do not cross the threshold of an armed attack and, therefore, remain a peacetime measure outside the Law Of Armed Conflict (LOAC). Outside the remit of the LOAC, any resulting harm from economic sanctions or cyber warfare, whether direct or collateral, is of little consequence unless it brings a humanitarian crisis.[4] As with "severe suffering," there is no exact definition of a humanitarian crisis except to say that it causes severe mental and/or physical suffering. But given the ambiguity of "severe" that definition is not too helpful and, as I will describe below, open to conflicting interpretations.

Nevertheless, the virtue of economic or cyber operations lies in their sub-threshold status. Stopping short of armed conflict, economic sanctions are a morally preferable step, short of war that nations should try to take before resorting to armed attacks. Therefore, we *encourage* aggrieved states

---

4 for example, United Nations Security Council resolution 661 (1990) detailing sanctions against Iraq restrict the use of commodities and funds "but not including supplies intended strictly for medical purposes, and, in humanitarian circumstances, foodstuffs..."

to wage economic warfare before turning to armed force when they face aggression. If war must be a last resort, then economic sanctions are the penultimate resort. In this vein, the subsequent discussions will consider the scope of cyber warfare as the penultimate measure before war, one we might similarly encourage rather than immediately restrain.

What are the Appropriate Goals of Sub-Threshold Cyber Warfare?

During armed conflict, the lawful use of armed force allows belligerents to kill enemy combatants directly and harm civilians collaterally to defend themselves against aggression. Permission "to kill," however, is to overstate things. The St. Petersburg Declaration (1868), one of the first expressions of humanitarian law, phrases things slightly differently:

- The only legitimate object which States should endeavour to accomplish during war is to weaken the military forces of the enemy;
- For this purpose, it is sufficient to disable the greatest possible number of men [that is, soldiers].[1]

Ultimately, war is not about killing, but about debilitating and disabling an enemy to neutralize a threat. During armed conflict, cyber operations may achieve these goals too. One obvious target for cyber operations is the military. A military target comprises those objects "which by their nature, location, purpose or use make an effective contribution to military action and whose partial or total destruction, capture or neutralization offers a definite military advantage (ICRC, Rule 8)."[2] The combined U.S./Israel Stuxnet attack on Iranian nuclear enrichment facilities is an excellent example of using cyber technologies to partially destroy or neutralize a militarily advantageous target. Dual-use targets are also a legitimate goal of cyber warfare. A dual-use target serves civilian and military populations and when the military attributes meet the conditions that define a military target, the target loses its immunity from attack. Dual-use targets might

---

[1] Declaration Renouncing the Use, in Time of War, of Explosive Projectiles Under 400 Grammes Weight. Saint Petersburg, 29 November / 11 December 1868. https://ihl-databases.icrc.org/ihl/full/declaration1868.
[2] ICRC (International Committee of the Red Cross). Rule 8, Customary IHL). https://ihl-databases.icrc.org/customary-ihl/eng/docs/v1_rul_rule8.

include many critical transportation, internet, or financial infrastructures that serve military forces and civilian populations.

Restrained Attacks on Civilians

Regardless of the target, the principle of noncombatant immunity and proportionality constrain military operations. Because civilians are innocent, that is, non-threatening, they do not lose their right to life. As such, international humanitarian law strictly prohibits *direct* attacks on civilian targets. Such attacks are grave breaches of international law and tantamount to war crimes. "Attack" usually means to threaten life and limb. But what happens when cyber operations do not threaten life and limb?

Collateral harm presents the same puzzle. Despite the principle of noncombatant immunity, civilian protections are not absolute. During war, belligerents may harm civilians *collaterally* when necessary to destroy or disable military and dual-use targets. However, the principle of proportionality only permits the killing of civilians when it is not excessive relative to the expected military benefit of an operation. Injury and loss of life are the metrics of excessive harm. If cyber operations cross the threshold of armed force, they are similarly constrained by proportionality. But if cyber operations remain below the threshold of war, there is no straightforward method to calculate proportionality. Military advantage continues to define the benefits of an operation, but what measures assess harm if no one is killed or injured? Are there then no limits to sub-threshold cyber warfare? Are civilians, then, legitimate targets?

Unrestrained Attacks on Civilians in Sub-Threshold Cyber Warfare

The challenges of cyber operations during armed conflict are the subject of another chapter in this volume. Here, I focus on sub-threshold cyber warfare. If we are unsure about extending noncombatant immunity during wartime cyber operations, we are equally uncertain about civilian protections from cyber attacks in peacetime. The problem is particularly acute when we think about operations directed against civilians that do not threaten life and limb, but hope to generate discomfort, inconvenience, or hardship. Here the goal is not to significantly harm the civilian population but to compel them to pressure their government to end their acts of aggression. In other words, sub-threshold attacks on civilians aim to indirectly weaken military forces by denying the government the political will to wage war. It's tricky because the causal line from hardship to pressure to compliance is pitted. There is no assurance that pressuring civilians will lead them to complain or defy their government, or that their opposition will affect government decisions. Nevertheless, direct attacks against the civilian population may constitute a legitimate goal when exercising force-short-of-war. This thought is nothing new; it has characterized economic warfare—sanctions and sieges—since biblical times. The result is to redefine the principle of proportionality and focus on the effectiveness and necessity of direct attacks against civilians. Comparing sub-threshold cyber warfare with economic warfare focuses the point more closely.

Cyber Warfare and Sanctions: Measures and Countermeasures for Utilizing Sub-Threshold Force

Cyber and economic warfare may supplement or supplant armed conflict. During full-scale war, each might try to weaken enemy forces by destroying military servers or blocking arms shipments in addition to striking one another's military targets. Think of the British blockade on Germany in World War I. The same blockade also targeted civilians, directly and collaterally, and led to 750,000 deaths. At the same time, cyber or economic warfare may intentionally target civilians to avoid or supplant armed conflict. This is sub-threshold warfare. Economic sanctions may deny discretionary assets that make life comfortable or block essential commodities to inflict severe hardship on the civilian population. The only constraint is a loosely formulated legal prohibition to prevent a humanitarian crisis. While sometimes a harsh byproduct of economic sanctions, humanitarian crises are not the obvious outcome of cyber operations.

In a typical cyber attack, hostile agents, whether state, non-state or unaffiliated, steal credit card information, interfere with the financial markets, threaten the destruction of critical infrastructures, disseminate false information about life-threatening attacks, compromise proprietary information and undermine the integrity of the electoral process. The consequences of these operations are not benign. Through simulated cyber attacks and natural experiments, political psychologists catalog significant stress, anxiety, and anger that accompany cyber attacks.

While similar emotional costs attend economic warfare, the ill-effects of sanctions and blockades may reach further to bring vast food insecurity, disease, poverty, unemployment, and death. In Gaza, Israel's blockade prevented arms shipments but also blocked building materials to deny Hamas the means to construct tunnels and fortifications. As residential and commercial construction suffered, unemployment, poverty, and unrest rose. In Iran, too, sanctions ravished the economy by shrinking oil production, devaluing the currency, and increasing the cost of living. Unemployment rose and inflation ran rampant. As public health deteriorates and poverty levels increase, the civilian population suffers greatly. But that is the point of sanctions. Still, and despite the suffering, few have suggested that the sanctions imposed on Gaza or Iran constitute a level of destruction or misery comparable to an armed attack that would justify a response tantamount to war. In Gaza, for example, UN and foreign humanitarian assistance help Gazans avoid an impending humanitarian crisis. Such was probably not the case in Iraq. Following the imposition of sanction after the First Gulf War, tens of thousands of Iraqis died from the indirect results of sanctions that brought severe food insecurity and decimated the healthcare system.

If sanctions and blockades in Gaza, Iran and Iraq represent a range of harm, some permissible and some not, then sub-threshold warfare, whether economic or cyber, has considerable room to maneuver. Short of an Iraq-like humanitarian crisis accompanied by widespread death and disability, starvation, and disease, cyber operations can cause considerable havoc.

Maneuvering in Sub-Threshold War Space

Keeping below the threshold and avoiding armed attacks, nations may use cyber operations to restrict the flow of war and dual-use materiel. Like economic sanctions, cyber operations aim to weaken enemy forces and/or impose hardship to undermine morale so that civilians compel their government to lay down their arms or comply with the attacker's demands. Economic sanctions, however, are notoriously slow and ineffective. In the worst cases, sanctions precipitate grave humanitarian crises that an authoritarian regime like Iraq or North Korea simply ignores. Can cyber attacks do any better? This is the test of effectiveness. If they cannot accomplish their goal, then cyber operations are no more permissible than sanctions. If they can achieve their aims, then cyber operations, like sanctions, must be proportionate. Before considering proportionality, however, let's examine two cases more closely.

Case 1: Cyber Aggression and an In-Kind Response

Consider the kind of scenario the Tallinn Manual 1.0 offers to guide feuding states A and B:

> State B initiates a cyber operation against an electrical generating facility at a dam in State A to coerce A into increasing the flow of water into a river running through States A and B. In response, State A may lawfully respond with proportionate countermeasures, such as cyber operations against State B's irrigation control system.

Note that cyber warfare aims not to disable or incapacitate an aggressor, but to attack civilian targets to compel the aggressor to obey international law and desist. Once State A backs off, no further operations are required or permitted.

Case 2: Aggressive, Defensive, and Retaliatory Cyber Operations

In addition to firewalls or decoy systems to block or confuse hackers, commercial and military servers in State A install active computer defenses (ACD) that initiate "hack backs" in response to unlawful cyber aggression. Hack backs are voracious. They do not merely prevent or disable the purported unlawful cyber attack, but reach beyond an offending network to breach the broader networks, programs, data, and servers of hostile agents to destroy malware and the threatening servers more expansively in State B. Such hack backs may occur automatically or play out over time in a series of well-planned retaliatory cyber strikes.

Common to these two cases is the idea of unlawful cyber aggression. In Case 1, cyber aggression means to disable a dam, thereby contravening treaties between the sides. Case 2 does not specify the nature of the unlawful aggression. Some cases might be clear and include the theft or destruction of data, financial assets, or other property. Other forms of aggression are less apparent. Arguing that "the internet is not indispensable to the survival of the civilian population (§81.5)," the Tallinn Manual 1.0 shakes off cyber operations that block email or internet services (§30.12), involve "mere economic coercion" (§11.2), or intend solely to undermine confidence in a government or economy (§11.3). For Tallinn, these acts are insufficiently threatening to rise to the level of an armed attack or justify the use of force in response. But they may be unlawful. Short of war, they remain the purview of sub-threshold cyber operations and may draw counter attacks.

One may, of course, argue this point. Attacks on the internet may very well be an act of war. The point is that cyber aggression remains fluid even when it does not cause injury or loss of life. Some acts that appear benign today may constitute hostile, war-like behavior in the future. In Tallinn 2.0, for example, the panel of experts conceded that "some may categorize massive cyber operations that cripple the economy as a use of force, even though economic coercion is presumptively lawful (Rule 69 (10)." I will return to this point at the end of the chapter.

Despite the commonalities of the two cases, the target's reaction distinguishes one from the other. In Case 1, the target's (State A's) reaction is limited to force commensurate with the original infraction and ends with the aggressor's (State B's) compliance. In Case 2, on the other hand, retaliatory force is unchecked by the initial aggression. The attacker is not only aiming for compliance but deterrence. And deterrence requires considerable force. Nevertheless, sub-threshold warfare does *not* make room for a principle of proportionality that weighs military advantage against harm to civilians or leave room for deterrence.

Proportionality in Sub-Threshold Warfare

Proportionality in sub-threshold war differs significantly from its wartime, *in bello*, cousin. Legally enshrined in Additional Protocol I (API 51.5) *in bello* proportionality prohibits "an attack which may be expected to cause incidental loss of civilian life, injury to civilians, damage to civilian objects…which would be excessive in relation to the concrete and direct military advantage anticipated." *In bello* proportionality dictates a vague ratio between two incommensurate variables: civilian harm and military advantage. Neither readily admits of a common denominator. A typical solution is to translate military advantage into human lives (while ignoring the effects of an attack on military capabilities, morale, or deterrence) and simply compare enemy civilian lives lost with compatriot civilian lives saved. When the former is somehow excessive compared to the latter, an attack is disproportionate.

In contrast, there is another kind of proportionality best characterized as tit-for-tat. This kind of proportionality is sometimes called a "countermeasure." Common to trade disputes and applicable to sub-threshold cyber warfare, countermeasures permit only an *equivalent* response that is no more severe than the original affront. Countermeasures are a kind of unarmed reprisal. Armed reprisals against noncombatants are long-discredited, but the underlying idea is the same: to compel a lawbreaker to comply with the law by exacting an equal measure of harm. Reprisals are guided by the magnitude of the initial infraction, not the more expansive goal of military victory that might easily demand far harsher measures than an equivalent response to a breach of law. While equivalence restrains countermeasures, *in bello* proportionality places no limits on the amount of military destruction that a belligerent may wreak. If no innocents die in the process, there are virtually no restraints on the use of force in war. Unlike *in bello* proportionality, countermeasures limit the destruction of *military* targets. There are two reasons for the narrow range of countermeasures. First, countermeasures hope to enforce the law and restore law-abiding behavior, not vanquish an adversary. Second, countermeasures endeavor to prevent escalation and the slide into armed conflict. On the other hand *in bello* proportionality assumes the full fury of war.

Countermeasures, then, are acts of "self-help" that usually include temporary and sometimes unlawful, coercive measures to force compliance following breaches of treaty, environmental damage, human rights violations, or aggression against third parties. Countermeasures include "non-forcible" measures or a response in-kind but preclude the "threat or use of force as embodied in the Charter of the United Nations."[1] In most cases, cyber countermeasures would enforce the legal regime governing cyber activities that prohibits hacking, computer fraud, espionage, and destruction or damage of data (see, for example, the U.S. Computer Fraud and Abuse Act, 18 U.S.C. 1030).[2]

Countermeasures hope to expect to enforce compliance and impose sufficient costs so that offenders do not profit from violating the law. Countermeasures usually translate into equivalent dollar costs, not comparable civilian deaths, which, by definition, do not characterize subthreshold cyber warfare. Such an attack would be *narrowly* proportionate because it brings similar kinds of harm to an enemy that does not exceed the harm caused by the initial attack. Returning to the two cases above, Case 1 exemplifies equivalent or "narrow" proportionality. Case 2, on the other hand, speaks to broader, *in bello* proportionately. The magnitude of the hack back does not rest on tit-for-tat, but on the goal of disabling an enemy so it can no longer pose a threat. As such, the retaliatory strike may be far more

---
1 International Law Commission, Draft Articles of Responsibility of States, 2001; Article 22.1; Article 50.3.
2 18 U.S. Code § 1030.Fraud and related activity in connection with computers.

forceful than the original attack. Doing so, however, often creates grounds for a vicious cycle of counter-reprisals. For these reasons, countermeasures emphasize their one-shot nature.

Liability in Sub-Threshold Warfare

Retaliatory cyber attacks to enforce the law conform to the self-help logic of reprisals but can be intensely problematic even when narrowly proportionate. If, in Case 1, the targets (a dam and irrigation system) are equivalent, it is only because each is a water facility. But there is no necessary equivalence of aims, harm, or liability to attack. In Case 1, we don't know whether A or B is at fault. The attempt to divert the headwaters of a shared river may be part of a long-standing feud over hydroelectric power. Perhaps the dam State A built contravened a treaty, and State B's attack is the permissible self-help countermeasure. Assessing blame and liability are common problems of international disputes, and frequently animate charges and counter-charges of aggression. What we do know, however, is that the farmers in State B are innocent. Attacking their irrigation system only jeopardizes their livelihood.

Therefore, the moral basis for the countermeasures is weak unless, paradoxically, the conflict is an armed one and cyber operations supplement (rather than supplant) the war effort. In that case, a war is raging, and the chief requirement for attacking any target is to neutralize an emergent military threat, not coerce an adversary to behave lawfully. Civilian targets are off the table, but an attack on State A's dam or State B's irrigation system is permissible if either is a military or dual-use target. In which case, civilians may suffer collateral harm subject to broad, *in bello* proportionality. Collateral harm is unavoidable but no one thinks that civilians are liable to direct attack.

A sub-threshold operation, however, targets civilians *directly*. One must, therefore, wonder about liability. Are the innocent farmers of State B blameworthy in some way and, therefore, liable to a direct attack? Economic warfare does not ask about liability. Intentionally targeting civilians, sanctions and blockades only inquire after comparable harms. But maybe we should be asking this question, so that any decision to harm others requires a degree of culpability and, therefore, liability to attack. In short, the targets of a cyber attack must be presumed guilty of, or the beneficiary of, some infraction.

One place may be to look to the beneficiaries of diverting the water. Is the water intended to enhance the supply of electricity? If so, State B's electric grid may be a better target than the farmers' water supply because 1) State B targeted an electrical generator to kick off the dispute and 2) those potentially benefiting from State B's action now pay the price for State B's unlawful conduct when State A retaliates. If, on the other hand, the farmers in State B benefit from the additional water, then disrupting their irrigation system is in place. This conclusion is not an appeal to liability *per se* but to taking action that denies the intended targets the benefits they are poised to receive from unlawful conduct when possible.

I stress "when possible," because liability is still, at best, a secondary consideration. There is no clearly defined class of liable actors in sub-threshold warfare as there is during war. There are only the government and its citizens. Faced with a neighbor's law-violating conduct, the first rule should be to exercise effective, equally harmful means to bring compliance. Should multiple targets present themselves, a further criterion would demand the search for liable targets or those benefiting from the initial infraction.

Liability, Proportionality, and Attribution

In the cases just described, attribution is not at issue. States A and State B know each other. But we know how the lack of attribution may easily plague cyber warfare as unknown agents, acting at the behest of a recognized state or non-state actor, launch aggressive cyber operations. Interestingly, it seems that as confidence in attribution decreases, certainty about intentions may increase. We may not know the identity of Ransomware attackers, data thieves, ballot box vandals, or instigators of inflammatory misinformation, but their aggressive, unlawful intentions are clear. Whoever they are, their liability to a proportionate counterattack is not in question.

Nevertheless, the countermeasure analogy is inexact. Countermeasures work best when the target can pinpoint the aggressor's culpability and liability, and assess the damage his attacks cause. Judgments about liability, culpability, and harm are possible when attribution is uncertain. We can identify harmful, malicious cyber operations. Missing here, however, is the ability to gauge the harm the attacker will suffer from a countermeasure when we do not know the attacker's identity. It is not clear, for example, who will be affected by a powerful hack back and the damage they will suffer. The affected parties may include the perpetrators and innocent bystanders. There is no easy solution to this problem except to proceed cautiously, search carefully to identify the effects of retaliatory cyber measures, and cease operations after thwarting the initial aggression. Nevertheless, the tendency to push further toward a show of force, whether cyber or kinetic, to deter potential enemies is tempting. Doing so, however, jeopardizes the sub-threshold character of cyber warfare.

Sub-Threshold: For How Long?

Despite the attempts to draw a bright line between armed conflict and sub-threshold warfare, much depends on a vision of war that is changing. I noted earlier how The Tallinn Manual takes a somewhat sanguine view of cyber operations that block the internet or email or undermine confidence in a government. At least, they do not believe that these kinds of operations are tantamount to terrorism. This view might be dangerously naïve. Terrorism, whether perpetrated by states or non-states works by undermining daily life, not by killing large numbers of people. Terrorism breeds fear, anxiety, political extremism, dysfunction, and lack of confidence in government institutions. Targeting civilians no less directly than kinetic terrorism does, many forms of cyber warfare impose similar costs on the innocent.

Consider the kinds of cyber attacks we have experienced to date. These include operations that publicize confidential information, rob people of their financial assets or identity, disrupt social networks and internet communications to disable virtual communities, manipulate the media to spread false information, and attack the mechanisms of fair and free elections. Attacks like these can undermine human security, imperil civil liberties, and erode public confidence in government institutions. The effects are, in many ways, similar to mass casualty events and push at the limits of sub-threshold warfare. Thus, limited, equivalent responses to what I have defined as sub-threshold cyber warfare may be insufficient to disable or thwart those inflicting the sweeping damage some non-lethal cyber attacks can cause.

Targeted governments, their people, and businesses must consider these potential damages when weighing appropriate, defensive countermeasures. Thus, and in contrast to the Tallinn Manual's legal opinion, a hostile government or guerrilla organization that tweets to cause panic by "falsely indicating that a highly contagious and deadly disease is spreading through the population," (§36.3) or pursues cyber psychological operations to undermine confidence in a government or economy, bears substantial exposure to blistering reprisal. One has to ask whether disabling these perpetrators should permit more than an equivalent response in kind.

Questions like these bring us to think about destruction in terms of its effect on human security broadly construed. For example, Hannah Arendt describes war as the ever-present threat of death, a "bestial, desperate terror which, when confronted by real, present horror, inexorably paralyzes everything..."[1] These are the effects of armed conflict on civilians and soldiers as we commonly perceive them. In contrast, Jeremy Waldron describes a different kind of terror, "a government coerced by the loss of something it values very highly—indeed, something indispensable for its status as government—namely, the ability to command and mobilize a large civilian population." "By rendering or threatening to render the population mindless with terror," Waldron continues, "the intimidator deprives the target regime of something it needs, a population capable of rational choice."[2] However, there is no need that a population is rendered "mindless with terror" to undermine its rational decision making capability. And in fact, Waldron looks to something short of "bestial desperate panic" to include "a state or condition that governments cannot afford to let their populations fall into or languish in for long." The outcomes of such an occurrence include the "collapse of economic morale," feelings of insecurity, apprehension and the disruption of social intercourse and daily life. These are precisely the effects we can expect of increasingly sophisticated sub-threshold cyber operations even when they do not cause significant death or bodily injury.

So if armed conflict necessarily brings abject destruction, injury, and loss of life, then cyber warfare does not pass the threshold of war. But let's think a little out of the box. We can easily imagine how cyber operations can lay waste in ways no less devastating or dangerous than armed conflict by undermining well-being, morale, public trust, and governability. To accomplish this end, one need not commit horrific acts of murder. In modern society, it is enough to strike at the foundations of everyday life: the ballot box, media outlets, or proprietary information (privacy). Among these, cyber networks stand out. As it targets these networks, cyber warfare is looking more and more like armed conflict. As a result, cyber warfare and its governing rules are a moving target that requires constant attention and (re) evaluation.

---

1 Hannah Arendt, *The Origins of Totalitarianism*, New Edition. (New York: Harcourt, Brace, Jovanovich, 1973).
2 Jeremy Waldron, Terrorism and the Uses of Terror. *The Journal of Ethics*, 8:1 (2004), 21-23.

References

18 U.S. Code § 1030.Fraud and related activity in connection with computers. 18 U.S. Code § 1030.Fraud and related activity in connection with computers https://www.law.cornell.edu/uscode/text/18/1030.

Arendt, H. *The Origins of Totalitarianism*, New Edition, New York: Harcourt, Brace, Jovanovich, 1973.

Dinstein, Y. *War, Aggression and Self Defense*. 4th edn. Cambridge: Cambridge University Press, 2005.

ICRC (International Committee of the Red Cross). Rule 8, Customary IHL). https://ihl-databases.icrc.org/customary-ihl/eng/docs/v1_rul_rule8.

International Law Commission, *Draft Articles on Responsibility of States for Internationally Wrongful Acts*, November 2001, Supplement No. 10 (A/56/10), chp.IV.E.1, available at: https://www.refworld.org/docid/3ddb8f804.html [accessed 20 September 2020].

United Nations Security Council (UNSC). (1990). *Resolution 661: The Situation between Iraq and Kuwait*. U.N. Doc. S/RES/661. file:///C:/Users/user/Downloads/S_RES_661(1990)-EN.pdf.

Waldron, J. Terrorism and the Uses of Terror. *The Journal of Ethics*, 8:1 (2004), 5-35.

Walzer, M. *Just and Unjust Wars: A Moral Argument with Historical Illustrations*. New York: Basic Books, 2006 [1977].

3

THE RIGHTS OF THOSE TARGETED IN MILITARY CYBER OPERATIONS

Michael Skerker

This chapter will explore how human rights serve as a constraint on the military aspirations of service personnel, regardless of the weapon system used. Consider the following cases as reference points for the argument to follow.

Imagine military acts of violence in three different eras:

1240 C.E., the German countryside
1A. A knight stabs a farmer with a sword.
2A. Soldiers torch fields which supply crops to the combatants and noncombatants in a besieged castle.
3A. Soldiers fire catapults at a castle, foreseeing that some noncombatants will be killed alongside combatants.

1940, the German countryside
1B. A soldier shoots a farmer with a rifle.
2B. Airmen bomb a bridge which is used by both military and civilian vehicles.
3B. Airmen bomb a munitions factory, foreseeing that some bombs will miss and hit nearby apartments.

2020, Germany
1C. A foreign cyber warfare engineer hacks the control system for a hospitalized German farmer's respirator and turns it off, killing the patient.
2C. During a war between Germany and country X, cyber warfare engineers from X use a malware attack to knock German communications grid off-line, affecting both military and civilian communications.

3C. Undercover agents from country X drop malware-infected zip drives in a parking lot at the Frankfurt airport, anticipating that one will be picked up and used in a computer in a co-located Air Force installation, where the code will disable German air defenses. The agents also foresee that the zip drive could be inserted into a civilian air traffic control computer, disabling the air traffic control system.

Weapon systems change but morality remains the same. We can make judgments about these cases according to the terms of *jus in bello*, the Western tradition of just conduct in warfare that is drawn from medieval warrior codes, Christian natural law theory, and international law.[1] The actions in cases 1A-C, above, are immoral because a military actor is directly attacking a noncombatant. Assuming the farmer is non-threatening, the killings are instances of murder. In cases 2A-C, the attacks are on dual-use infrastructure used by both military and noncombatants. In cases 3A-C, a military target is attacked, but the actors foresee that noncombatants will be harmed as an inevitable side effect. Both latter types of attacks (2 and 3) are permissible if the military targets are necessary to the attacker's strategy; there is no alternative style of attack lacking the associated noncombatant harms or with fewer harms than the contemplated operation; and the amount of harm done to noncombatants as a side effect is proportional to the tactical good done in the operation.

Cyber operations provide military actors with new ways to frustrate their enemies' operations and as it happens, new ways to harm noncombatants. Any civilian system that uses the internet or computerized or robotic controls can in principle be disabled or negatively affected with remote cyber operations. In some cases, hackers can physically destroy civilian systems using cyber means or erase data from storage systems. Any such civilian system can also be destroyed with conventional weapons, but of course, what is tantalizing, or terrifying, about cyber operations, is that effects on par with kinetic attacks can be achieved from the other side of the globe nearly instantaneously, covertly, selectively, deniably, relatively cheaply, and temporarily. The final aspect gives the policymakers options of coercion below the threshold of war; for example, country A could power down a rival's electrical grid for an hour as a warning or temporarily disable an adversary's air defenses as part of a strategy aimed at affecting a negotiation's outcome.[2]

---
1 There are analogous codes in non-Western philosophical and theological traditions.
2 David Whetham, "Cyber Chevauchees," in eds. Fritz Allhoff, Adam Henschke, and Bradley Strawser, *Binary Bullets: The Ethics of Cyberwarfare* (Oxford: Oxford University Press, 2016).

Yet novel military means do not change the moral calculus service members have been called on to consider for centuries. As the above cases showed, the same moral reasoning applies to the use of different weapon systems because of the rights of those targeted.

Moral Foundations

In most cases, other people's rights act as a limit to one's aims, forcing one to make room for other people's aims. Rights delimit one's justifiable range of choices and actions. In rare cases, individuals' rights can be overridden when not doing so would have terrible consequences for large numbers of people. So, what are rights? Where do they come from and who has them? How ought they to limit the operations of cyber warfare engineers?

The idea of natural human rights is a very powerful one, growing in acceptance by people around the world, with many skeptics remaining to be converted. The idea of universal human rights presupposes moral equality: that all human beings, no matter where they live, no matter their wealth, intelligence, class, profession, and so on, are morally equal. Their morally permissible interests (more on this below), hopes, dreams, relationships, opinions, etc. count just as much as any one else's.

Scholars differ as to the source of this equality. Some say that people are equal in that they possess a quality called dignity, a special status that the great philosopher Immanuel Kant defined as a value that cannot be exchanged for anything else. In other words, we might pay for a tool or trade one tool for another, or offer a service to pay for a tool, but humans themselves should not be bought, sold, or traded. There is no dollar value to be placed on a person, because unlike a tool, often designed to do one specific thing (that has a certain value), humans can do millions of different things. One might only pay $20 for a hammer, because one knows exactly how one will use it and the value of that project, but one cannot hold a baby and know all the millions of things she will do in her life and their value.[3]

Religious believers might say that all humans have dignity because they are created in the image of God and so have an innate sacredness.[4] On account of this sacredness—this divine parentage—it is wrong to treat any person, no matter what he has done, in a disgraceful way. It would be an

---
3 Insurance actuaries do estimate how much a given person might *earn* over her lifetime, say, in cases of wrongful death civil suits, but this is not really an estimation of a person's total worth.
4 As one example, see Pope John Paul II, *Evangelium Vitae* (Boston: Pauline books, 1995).

insult to the Creator. Dignity is often talked about when people are treated in an undignified way: when they are humiliated or tortured, when they are reduced to an animalistic state. It is particularly those at the margins of society—the poor, refugees—and those who earn society's contempt, like criminals, drug addicts, and prisoners, who are vulnerable to being treated in an undignified way. It sometimes takes an especially morally thoughtful person to see these people's dignity, their innate worth as people—which was more obvious before bad luck and bad choices took their toll—as well as their capacity for redemption. On a broader level, as Mahatma Gandhi famously said, "The true measure of any society can be found in how it treats its most vulnerable members."

Other scholars see moral equality as rooted in people's capacity for autonomy. Autonomy means self-government, the ability to make one's own decisions and chart a course in life. Even though some people seem not to express their autonomy very much, instead just following the crowd, while others live in communities or countries where their autonomy is denied with restrictive laws, the idea is that everyone has the ability to control their immediate desires; impose a moral rule on themselves; and choose an action independent of the frantic tug of desires. For example, Joe takes out his just-purchased burger and sets it next to him on a park bench. A passing dog might jump and grab his burger. A person, even if a little hungry, would not, because she can control her instinct to grab the tasty-looking sandwich and making her actions conform to a self-imposed rule "respect other people's property" or "eat healthy food." Having controlled her own instincts and moderated her behavior in order to defer to other people's rights, she now has the space to make a free decision about what to do for lunch. Had she just grabbed the burger, she would have been unable to act on her rational desire to eat healthy food and also would have gotten into trouble for theft, further frustrating her aim to get a healthy lunch. She would be less free than someone who exercises self-control in that she is a slave to her desires.

What is interesting, is that it seems every mentally normal adult can engage in this kind of complicated act of self-control and self-direction. The moral theories that see autonomy as the distinctively human trait think this ability is so unique and special that it should always be honored, even if a person is not making many independent decisions or is not making good decisions.

Dignity, autonomy, or both can be seen as the foundation for moral equality, regardless of a person's nationality, class, race, profession, gender, etc. All people are worthy of a basic level of respect, deference, and care, regardless of their value to a particular person or society and regardless of what they have done.

Dignity and/or autonomy is the foundation for human rights. Rights can be thought of as expressions of autonomy in specific areas of thought, speech, or action. If autonomy is a type of freedom, rights delineate how far one's freedom to speak, act, believe, use property, marry, worship, form friendships, etc., extend against a backdrop of moral equality. All adults start out, at least, with the same set of rights, to be enjoyed to the same extent. A right gives one the freedom to act within the realm of operation the right identifies. A right of free speech means one is free to speak one's mind; a right to marry means one is free to marry the partner of one's choice; a right to private property means the freedom to use one's property as one sees fit, and so on.

More specifically, a right, say to private property, means one may use one's property as one wishes: destroy it, sell it, give it away, or lock it up. One can demand others defer to one's own decisions in this realm; one can use proportionate levels of force to defend one's right; and one can demand, and expect, help in defending one's right.

Since everyone is morally equal, everyone has the same set of rights, to the same extent. Put another way, A's rights stop where B's start. So, one can use one's property as one wants, but not if it violates other people's rights. One can take practice swings with one's baseball bat, but not on a crowded subway. One can listen to the music of one's choice on one's own radio, but not so loudly that others cannot listen to their own music. One can plant crops on one's own land, but not on someone else's land.

Relevant to military operations, foreign noncombatants have the same set of natural human rights that one's domestic neighbors enjoy. Again, it may be the case that an oppressive foreign government does not permit people to worship freely; read what they want; or criticize the government. It may be the case that many people in a foreign culture think it is permissible for men to beat or rape their wives or for members of the majority religion to enslave people of a minority religion. On the view of human rights theory, these practices are wrong. We can articulate the sort of reasons outlined above for why these practices are wrong. Human rights are embedded in human nature, rooted in dignity and/or autonomy, regardless of whether or not a culture or a government recognizes all of them for all people. It may also be the case that the objectionable practice is at odds with other values of the local culture, which is to say, wrong according to the lights of the moral system that is dominant in that culture.

It follows that just because someone lives in a different country, it is not the case that she lacks a right to life, health, security, and privacy. Now it is true, that as a matter of fact, cyber warfare engineers have the technical means to threaten or violate these rights with relative ease, potentially without reliable attribution. Yet just as it is a violation of her rights if she is shot by a soldier; if her local hospital is bombed; or if a foreign agent reads her diary; so too are her rights violated if a foreign actor does these sorts of things in a cyber operation, be it covert or overt.

It is worth considering the right to privacy at greater length since it is especially easy to violate electronically and its violation may not seem as problematic as violations of some other rights. A moral right to (mental) privacy is important to express and defend one's autonomy. This right gives one power to protect one's thought and deliberation, without which, one would not have ideas of one's own to autonomously pursue. Since knowledge of others' thoughts, values, memories, opinions, and self-knowledge (for example of past actions) can give one power over the target, a right to privacy includes the power to control the release of this kind of personal information. It is not difficult to comprehend what a paralyzing effect it would have on one's life if one feared that all one's communications were being monitored by state agents and that anything one wrote or posted during one's entire digital life was also exposed.[5]

Waivers, Forfeiture, and Double Effect

Yet just as is the case with kinetic operations, it is not the case that service personnel are prohibited from actions that *risk* harm to noncombatants. To clarify this point, we first need to discuss how people can give up rights that would otherwise protect them from harm. A person can *waive* certain rights for a period of time through express consent, as when a boxer agrees to being hit in certain ways for the duration of a bout. As his opponent similarly waives certain rights, boxer A may attempt to strike his opponent, but is not wronged if his opponent hits him. However, the waiver terminates at the end of the bout. I would argue that service personnel waive a right against being attacked in certain ways by future conventional combatant

---

[5] For further discussion, see Michael Skerker, *An Ethics of Interrogation* (Chicago: The University of Chicago Press, 2010), ch. 3.

opponents when they enlist. This is to say, in a future war against another state or state-like entity, conventional enemy troops do not violate the rights of service personnel when they try to kill them. Both sides are permissibly serving their state's national security interests in a violent trade. Service personnel *do* violate their conventional enemies' rights if they violate the rules of war, such as by killing troops who have surrendered or using chemical weapons.

People can also involuntarily *forfeit* certain rights by *trying* to violate others' rights. An idea associated with autonomy is that all people have the capacity to restrain their instincts in order to defer to other's rights. Mentally normal adults can resist stealing someone else's food even if it smells very good because they know stealing is wrong. Yet if a person fails to restrain himself and say, tries to steal, others can do the restraining for him and, if need be, physically stop him from stealing. Normally, one cannot physically impede others, but in this case, the thief has forfeited his right to bodily autonomy (doing what he wants with his body) because he has failed to responsibly control his own bodily movements.

Such forfeiture is limited and temporary: one forfeits temporary enjoyment of the rights that need to be curtailed to stop one's rights violations until the just status quo ante has been restored. So, one can physically restrain a thief, but one cannot rape him; read his diary; or prevent him from voting. A mugging victim can use physical force to protect her property and person, but cannot prevent her assailant from marrying or worshipping as he wishes. Someone being unjustly attacked with lethal force can defend herself with lethal force if necessary, but cannot use lethal defensive force if someone is only insulting her. Further, one cannot hurt an attacker after the fact, say a month later, when one sees one's mugger sitting on a park bench. One might argue that the attacker deserves punishment or should be harmed to deter him from further aggression. This might be true, but many argue that in ordinary society, the aggrieved party is not the one who should dish out deterrents or punishments. Rather, a neutral third party who can better make a fair judgment about the attacker's desert should do so.

Finally, it can sometimes be permissible to harm innocent people who have neither waived nor forfeited their rights, justified by an argument known as the doctrine of double effect. The argument was developed to address situations where the actor is trying to bring about a great good, such as defending others' rights, but cannot achieve this good without, at the same time, harming other innocent people. Double effect is often used as a way of justifying kinetic situations such as the "B" and "C" cases in the above thought experiments. The Air Force wants to bomb a legitimate military target, say, a missile silo. Doing so will save at least X number of allied lives, but there is a civilian home within the blast radius. Normally, bombing a missile silo is permissible in war, so the question here is whether the action is still permissible under circumstances where there will be a *simultaneous and inseparable good and bad effect* as a result of one's normally permissible action: destroying the silo and destroying the home. When the bombs land, the silo and the home will be destroyed virtually simultaneously. There does not seem to be any way to destroy the silo without the negative side effect. Are the airmen heroes for destroying the silo or villains for murdering civilians?

There are different construals of the doctrine of double effect. One form is as follows. A militarily necessary action with simultaneous and inseparable good and bad effects is permissible if and only if:

a. The intended action must not be inherently immoral (for example rights violating) like murder, rape, enslavement, assault, torture, etc. [murder and assault being defined as intentional actions against innocent persons].
b. The bad effect must be unavoidable if the good effect is to be achieved. If there were some other course of action that could bring about the good effect without the bad effect (for example a ground assault on the silo), then the action would not be justified by the doctrine of double effect.
c. The bad effect is not a means to the good one. In other words, destroying the home is not causally necessary to destroy the silo. The silo could still be destroyed if no home was nearby. If the airmen hit the silo and the home survived, the airmen would not have to circle around and attack the home in order to achieve their mission.
d. The bad effect is proportional to the good effect (for example the good effect is roughly equal to or more significant than the bad effect).

This attack on a missile silo is potentially justifiable through the doctrine of double effect despite foreseeable harm to civilians. Bombing a silo in war time is normally permissible. Under certain tactical conditions, there may be no other way to destroy the silo without attacking it from the air. It may be necessary to destroy the silo immediately. Targeteers and weaponeers advise that there is no way to destroy the silo without also destroying the house as well. Even if people in the house are warned, there is no guarantee that they will heed the warning. The destruction of the house is not a means to destroying the silo—it is not the case that the silo's destruction depends on the house exploding. Finally, if the number of lives saved by the missile silo being destroyed outnumbers the number of civilians killed in the house and the attacking force does not think the resulting civilian deaths will be used for significant propaganda effect, the attack may be proportionally justifiable.

The Potential Targets of Cyber Operations

It has already been argued that changing weapon systems does not change the moral demands on those operating the systems. They still need to respect the rights of those potentially affected by the weapons' operations. We can now consider potential targets of cyber operations in order to clarify if they have waived or forfeited any rights or, failing that, if cyber operations can proceed, even at the risk of harming noncombatants, justified through the doctrine of double effect. We will consider the rights of noncombatants, service personnel, and unprivileged irregular militants. The last group can include members of organized sub-state armed groups including some insurgent groups attempting to overthrow a particular state, colonial regime, or foreign occupation. Rebellions of these sorts can be morally permissible if groups wear uniforms, obey a unified command, and obey the laws of war. Such militants have the moral profile of conventional combatants even though they do not have a state yet or anymore to defend. Irregular militant groups are *unprivileged* if they lack these characteristics and thereby, have more in common with criminal groups. Other militant groups that are usually unprivileged seek to foment transnational political movements through violent means. These types of groups, including al-Qaeda and ISIS are sometimes referred to a terrorist groups, (though this somewhat confusingly identifies a group with the tactics it may happen to sometimes use).[6]

Adversary service members or privileged irregular combatants waive the rights necessary for their opponents to halt their tactical actions

---

6 See *An Ethics of Interrogation* (Chicago: The University of Chicago Press, 2010), ch. 6.

THE RIGHTS OF THOSE TARGETED 53

or remove them from the battlefield.[7] It therefore does not wrong them in wartime to target them with cyber means used to disrupt their intelligence, surveillance, and reconnaissance (ISR) efforts; interrupt their tactical communications; disable or corrupt their computer systems; or disable their materiél even if there are potentially lethal effects on the operators (for example disabling the guidance system of aircraft). All this, provided that the operators adhere to the morality of *jus in bello* (described in an earlier chapter) including necessity, discrimination, proportionality, and a prohibition of unnecessary suffering. Cyber operations may often be morally preferable to kinetic operations to the same end if, for example, ISR technology, computer networks, and vehicles can be disabled without the kinetic effects that usually kill or maim their human operators.

Conventional combatants and privileged irregulars can also permissibly attempt to defend their networks; hack back; and go on offensive cyber operations targeting their enemy's computers, networks, communications, and other networked technology in return. As with boxers, neither side is wronged by the other's counter-force cyber operations.

Unprivileged irregulars can be directly targeted in all the same ways as conventional combatants though on account of a different justification. Unprivileged irregulars are bent on rights violations and so forfeit their rights to the technology, networks, and data they need to carry out, fund, and conceal their operations. For example, cyber warfare engineers can attempt to shut down ISIS Twitter accounts; hack into their servers and computer networks; and erase their data. Militant groups have no right to defend their networks or go on offensive cyber operations. In a similar fashion, armed bank robbers do not have a right to self-defense when they are confronted by bank guards.[8]

Ordinarily, noncombatants cannot be directly targeted with cyber operations to disable their technology; intercept or disrupt their communications; erase their data; or physically hurt them. As in conventional military operations, cyber operations can target dual-use infrastructure

---

7 It should be noted that the question of whether personnel on both the just and unjust sides of the war must adhere to the same moral standards has been contested by philosophers in the last 15 years. While the egalitarian ethic is reflective of international law and the historic consensus among moral philosophers for the last three centuries or so, "revisionist" scholars tend to argue that personnel on the just side of the conflict do not waive their rights and are wronged if they are attacked in the course of an unjust war pressed against their country. I argue against this view in *The Moral Status of Combatants: A New Theory of Just War* (London: Routledge Press, 2020).
8 Michael Walzer, *Just and Unjust Wars*, fifth ed. (New York: Basic Books, 2015), 127.

like airport control systems used by both civilian and military aviation or communication networks used by both provided the operation can pass the doctrine of double effect. Again, this philosophical tool is relevant to cases where one's action will have simultaneous and inseparable good and bad effects. It would not be controversial for service personnel to target an enemy's military communication grid if it were separated from the civilian grid. The attack becomes morally fraught because of the simultaneous and inseparable effect on civilians in the case when both the military and civilians depend on the same network. In this case, applying the doctrine of double effect, an attack on a communications network used by the military in wartime is a generically permissible action. The attackers would need to consider if the bad effect on noncombatants was unavoidable. The effect is physically unavoidable if both the military and civilians use the same network and there is no alternate civilian network. The attackers would need to consider if this attack needs to proceed now—if there is no other way to achieve the desired tactical advantage than by disabling this network. For example, the attack should not proceed if it was known that the military have a back up communications network; then the bad effect on noncombatants would be wrought for little purpose. The attack would also fail double effect if disabling a separate civilian network was causally necessary to disable the military network. For example, say a military network was ably defended with firewalls, but servers supporting the military network were co-located in the same facility as servers serving the communications network. A malware attack on the poorly protected civilian servers could cause a power surge that would also short out the military servers. In this operation, the bad effect on the civilian network would be a means to the good effect of disabling the military network. Finally, it has to be considered if the good effect of disabling the military network is worth disabling the civilian network. The "easiest" way to accomplish this end is to compare like effects: lives. The attacker would need to consider how many allied lives would be saved by disabling the adversary communication network and how many foreign noncombatant lives risked if the communications network fails. How many people would die because no one can call ambulances or no hospitals can share medical records, etc? Would the attacker's forces be able to achieve tactical dominance in a particular theater in the time it would take the adversary to repair its networks? Would that victory lead to the end of the conflict? The attacker might consider other effects as well. For example, if the disabled communications network would not make a difference in terms of lives lost or saved in combat (perhaps the armies are not in range of one another) or in civilian areas, the amount of economic losses and general panic caused amongst noncombatants might be enough on their own to render the attack disproportionate and therefore, impermissible by the lights of the doctrine of double effect.

The same sort of analysis can be applied to case 3C, from above, where the target is not dual-use infrastructure, but where one perceives that the attacker's action will set into motion positive and negative effects. A social engineering attack could lead to at least two morally significant downstream effects: disabling the enemy's air defenses, but also infecting the civilian air traffic control systems. A social engineering operation is not inherently immoral, even though it relies on deception and trickery for its success. Engaging in ruses to trick the enemy is permissible. As philosopher Sissela Bok argues, if the enemy waives his right to being attacked with lethal force, he also waives his rights against lesser forms of rights-infringing behavior such as trickery. This social engineering operation also passes the third element of double effect in that disabling the civilian air traffic control is not a means to disabling the military air defense system, but rather a pure side effect. However, it is doubtful that the operation passes the second element of unavoidability. Sprinkling infected zip drives throughout a parking lot in the hope that an airman picks one up and then uses it for official business is a very imprecise (even if sometimes successful) operation. There are probably other means of disabling the air defenses through cyber means or in more discriminately targeting the air defenses without raising such a high risk of infecting the civilian infrastructure. Considering the fourth element of double effect, the proportionality element, there are significant effects on either side of the ledger that must be considered. Disabling the air defenses of a hostile adversary has massive tactical consequences, saving the lives of large numbers of allied aviators, but disabling an airport's air traffic control system when jets are inbound could be calamitous. This operation is especially reckless, because the infection of the civilian air traffic control system could occur days after the air defenses were disabled, when a separate person pockets an infected zip drive and uses it in his airport office. Responsible attackers would need to find a different way, more discriminate, more focused, and more controllable.

In this chapter, we have considered the human rights that serve as brakes on the interests of military actors. Even if cyber means provide new more subtle ways of achieving military objectives, the same moral limits rooted in human rights obtain.

References

Pope John Paul II, *Evangelium Vitae* (Boston: Pauline books, 1995).
Skerker, M. *An Ethics of Interrogation* (Chicago: The University of Chicago Press, 2010).
Skerker, M. *The Moral Status of Combatants: A New Theory of Just War* (London: Routledge Press, 2020).
Walzer, M. *Just and Unjust Wars*, fifth ed. (New York: Basic Books, 2015), 127.
Whetham, D. "Cyber Chevauchees," in eds. Fritz Allhoff, Adam Henschke, and Bradley Strawser, *Binary Bullets: The Ethics of Cyberwarfare* (Oxford: Oxford University Press, 2016).

PART TWO

4

CYBER WARFARE AND CONVENTIONAL MILITARY OPERATIONS

Richard Schoonhoven

On 6 September 2007, a group of Israeli attack aircraft crossed into Syrian airspace and bombed a suspected clandestine nuclear reactor in the Deir ez-Zor region of Syria. The site was destroyed, and the Israeli aircraft (and a team of commandos apparently on the ground to laser designate the target) returned to Israel without further incident, and without suffering any casualties. The attack was facilitated by Israel's sophisticated electronic warfare capabilities, which allowed the Israelis to feed the Syrians a "false sky" picture for the duration of the operation. Basically, the Israeli aircraft went undetected, the Syrian air defense systems reporting everything as normal.[1]

In itself, the incident would seem to raise few ethical issues (over and above the justification of the strike in the first place, of course). Much of the ink that has been spilled about "cyber war" would seem to be beside the point here, if we restrict ourselves to the use of cyber—advanced computing resources and capabilities—as an adjunct to, or in conjunction with, traditional "kinetic" war, as we do in this entry. The questions discussed most prominently in the literature center on the notion of cyber war *per se*: Could there ever be a purely cyber war? When does a cyber-attack constitute an act of war? When might such an attack justify a kinetic response? Do the normal rules of war (International Humanitarian Law (IHL)/Law of Armed Conflict (LOAC) for example) apply to activities purely in the cyber realm?

---

1 Peter W. Singer and Allan Friedman. *Cybersecurity and Cyberwar: What Everyone Needs to Know*. (Oxford: Oxford University Press, 2014). Also cf. https://en.wikipedia.org/wiki/Operation_Outside_the_Box. The attack is usually referred to as *Operation Orchard*.

But even if these are open questions, it might be thought that since IHL/LOAC applies to all conventional armed conflicts, it ought to apply equally to any cyber activities that are part of those conflicts. That is, if the cyber realm is considered something separate and apart from warfare, there might be questions about the applicability of IHL, or of the Just War Tradition more generally. But to the extent that cyber is just one means among many in an otherwise "conventional" conflict, it might be thought that the standard rules unproblematically apply.[2] This is especially true given that many of the activities to which cyber capabilities particularly lend themselves—espionage, sabotage, communications disruption, signals intelligence (SIGINT)—have perhaps always been part of war and warfighting. Militaries have long attempted to break the enemy's lines of communication: messengers have been intercepted,[3] radio signals have been jammed, telegraph wires have been cut. In WWII, German sharpshooters were deployed to shoot down any homing pigeons that might be carrying messages back to Britain. And communications have not just been prevented, they have also been altered. A paper message, if intercepted, could be re-written, and apparently pigeons weren't just killed, but were also "hacked:" captured "enemy" pigeons might be given fake messages to carry.[4] Espionage too, has long been a part of warfare. Forward observers have been posted, spies have been sent behind enemy lines, and communications have been tapped, all to gain a strategic advantage through greater knowledge of the enemies plans, movements, and intentions. This too is one of the areas in which cyber excels: if one can infiltrate an enemy's command and control computer systems, one might gain extensive knowledge of his capabilities, etc. In the extreme, one might even be able to take control of crucial systems, perhaps turning them against the enemy himself. Even if such dramatic turn-abouts are not possible, on the modern battlefield there are any number of systems and pieces of equipment that might have their functionality compromised—electronic

---

[2] Certainly, this seems to be the opinion of experts, at least as far as LOAC/IHL is concerned: "It is clear that when cyberoperations accompany kinetic hostilities qualifying as either international or noninternational armed conflict [...] IHL applies fully to those operations." Michael N.Schmitt and Liis Vihul. "The Emergence of International Legal Norms for Cyberconflict." In Fritz Allhoff, Adam Henschke, and Bradley J. Strawser (eds.). *Binary Bullets: The Ethics of Cyberwarfare*. Oxford: Oxford University Press, 2016).

[3] A rather charming example: the Modern Olympic Pentathlon is reportedly constructed around the idea of a messenger trying to deliver a message from behind enemy lines in the Napoleonic era. He had to ride a horse, run, swim, fence a duel, and shoot his way clear to deliver his message.

[4] "Birds of War", Historynet, Jan 2019, https://www.historynet.com/birds-of-war.htm.

sabotage—rendering them inert and useless. In the extreme, actual "kinetic" damage might be done; Stuxnet (to be discussed below) was proof-of-concept for this.

All of this is just to say, then, that concealing one's actions and disrupting enemy communications is nothing new, and this particular attack by Israel would seem to be a model of the sort of precision-strike capabilities that cyber warfare enables, limiting damage and loss of life. It seems safe to say that never again will a war be fought without a robust cyber (warfare) component (at least not unless one side is extremely technologically primitive, in which case the overmatch is likely to be so utter that the conflict might hardly count as a war), and to the extent that this allows for greater precision and (therefore) discrimination, the development is to be welcomed.

So, we might wonder what all the fuss is about. But not everything here goes in the "plus" column. I will argue that the inclusion of cyber activities, even in otherwise standard warfare, does introduce ethical problems and complexities, exacerbating old concerns and perhaps introducing genuinely new ones. In any case, the question is worth investigating. The traditional areas of concern in the Just War Tradition are *Jus ad Bellum*, the justice of the resort to war in the first place; *Jus in Bello*, the justice of one's conduct in a war, given that fighting has broken out; and, more recently, *Jus post Bellum*, justice after war, dealing with questions of reparation, reconstruction, etc. (this last has been particularly urged by Brian Orend). Others have tried to add others, but these would seem to be the big three that have garnered the largest consensus, at least as regards their importance, if not their content. We will discuss them in turn.

Jus ad Bellum

In the Operation Orchard raid, not a single Israeli life—not even a single Israeli aircraft—was lost. To the extent that advanced cyber capabilities hold out the promise of greatly reduced casualties on one's own side, might not the threshold for the resort to war be lowered? There can be little doubt, after all, that the Israeli willingness to launch the strike in the first place was increased by the promise of security that their cyber capabilities afforded. Thus, we encounter a version of what has become known as the Threshold Problem, where war becomes more and more tempting, and the threshold to declaring it commensurately lower. Although one of the traditional *ad Bellum* requirements has always been that war should be a last resort, it can be difficult to state precisely what this actually requires. At least until the bullets start flying—and even then, if one is so inclined—one can always try one more diplomatic overture, one more economic sanction, one more international appeal.[5] In practice, obviously, one of the great brakes on politicians declaring war, or on getting involved in foreign misadventures more generally, is the threat of young men and women coming home in body bags. Politicians—people in general, sadly—are often much more concerned about their own dead—even the deaths of their own combatants—than they are about deaths among the enemy. To the extent, then, that a war can be launched with very little risk to one's own side—with very few expected casualties—then the bar to launching wars gets lower and lower, and the resort to war easier and easier—politically, if not ethically. Thus, political leaders might be tempted to begin wars that they otherwise would not. Even if these wars are (otherwise) justified, the proliferation of wars is problematic, in that they eventuate in the deaths of people who would not otherwise have died. Without the option of a "low cost" war, perhaps the relevant leaders would find other, more peaceful, means to achieve their goals, or would even let those goals go unrealized (whether this would be worse or not, of course, depends on the details of the particular case).[6]

A similar worry can be expressed about, and through, the requirement of proportionality. To the extent that cyber capabilities genuinely make war less destructive, they change the proportionality calculation. If we think of a war that is otherwise fully justified—a war of defense against naked aggression, for example—then this is of course to be welcomed. A truly just war fought at a lower cost is a good thing. But again, the worry is that political leaders will be lured by the apparent lower cost of such wars into fighting wars that are less than fully justified otherwise, or wars that don't strictly need to be fought.

---

5 Consider the controversy over this requirement in regard to the 2003 U.S.-led invasion of Iraq. One way of framing one of the issues in the controversy over whether the U.S. was justified in taking military action against Iraq was whether the requirement of last resort had been met. President Bush argued that it had been (although in fairness he also did somewhat change the rules of the game by moving toward a notion of preventive war), while many of the U.S.'s allies argued that it had not. See for example the Chilcot Inquiry in "Iraq Inquiry", Wikipedia https://en.wikipedia.org/wiki/Iraq_Inquiry.

6 Of course, this worry is not unique to cyber: it threatens whenever there is a radical power asymmetry—but especially a technological asymmetry—between two sides. Drones, human enhancement, etc., might all raise similar issues.

In any case, both the proportionality calculation and the reasonable chance of success requirement of the *Jus ad Bellum* are likely to be much harder to figure out in a war that relies heavily on cyber capabilities. Proportionality, whether a war will be worth it in the long run, is always difficult to figure out—as Michael Walzer points out, we are much better at recognizing gross disproportionality—and the reasonable chance of success criterion can really only require that losing isn't a foregone conclusion. But at least in conventional wars, one usually has at least some idea of the enemy's strengths and capabilities: troop strength, number and type of aircraft and munitions, that sort of thing. But this might not be the case in the cyber realm. It is apparently possible for a nation's critical systems to be compromised for an extended period without that nation even realizing it—witness the Solar Winds hack into numerous U.S. Government networks that went undetected for months.[7] If one cannot even be certain of the security and reliability of one's own systems, then it becomes extremely difficult to calculate the likely outcomes of something as chaotic as war.

Jus in Bello

Here too the news is both bad and good. On the one hand, cyber has the potential to be the most surgically precise, and therefore discriminate, weapon the world has ever known. Stuxnet is the poster child here. Stuxnet was a virus, presumably engineered by the U.S. and Israel, although neither has ever officially acknowledged responsibility, that was used to attack the Iranian uranium enrichment program. If it found itself on a computer controlling a Supervisory Control and Data Acquisition (SCADA) system manufactured by the Siemens corporation, the kind used by the Iranians to run the centrifuges used to obtain weapons-grade uranium (U^{235}), it did two things. First, it caused the centrifuges randomly to speed up and slow down. These are extremely precise, and therefore delicate, machines, and such variations in speed eventually lead to catastrophic failure. At the same time that the virus was literally causing the centrifuges to tear themselves apart, however, it was also sending diagnostic signals back to the centrifuge operators indicating that all was well and the system was functioning

---

7 See, for example, "As Understanding of Russian Hacking Grows, So Does Alarm", The New York Times, Jan 2021, https://www.nytimes.com/2021/01/02/us/politics/russian-hacking-government.html.

normally. Thus, the problem went undetected until considerable damage was done, with the Iranians trying to figure out why they were experiencing so many centrifuge failures. By most estimates, Stuxnet set the Iranian nuclear program back by at least a year.[8]

If nothing else, Stuxnet was proof-of-concept: first, that a purely cyber operation could have physical, kinetic effects in the "real world;" and second, that such a weapon could be exquisitely finely targeted. Stuxnet was designed to target only SCADA systems manufactured by a specific company used for a very specific function, essentially a unique identifier. If the virus found itself on other computers[9]—computers not used in the refinement of uranium, essentially—it basically did no harm.[10] It might have occupied a bit of space on the computer's hard-drive, but that's about it. It remains unclear precisely how the virus was introduced to the Iranian computers. The system apparently was air-gapped, and so the virus might have been introduced, intentionally or otherwise, by someone inserting a thumb-drive or some such. Thus, the virus may have been intended only to penetrate the targeted computers. But the point is that even though it in some sense missed its target—or better, hit many other targets in addition—it did no harm to these other machines. Indeed, the virus could presumably have been intentionally distributed quite widely, without doing any more damage than it was specifically designed to do.[11] This has led some to call Stuxnet the most ethical weapon ever deployed, and to the extent that cyber means allow nations to tailor their weapons so precisely, they hold out the hope of a considerable increase in discrimination and proportionality.[12]

Of course, Stuxnet was a "purely" cyber attack, with no use of conventional, kinetic weapons, and our focus here is on cyber weapons as part of, as an adjunct to, conventional warfare. But the same considerations still apply. Imagine a cyber weapon that attacked *only* an enemy's

---

[8] Peter W. Singer and Allan Friedman. *Cybersecurity and Cyberwar: What Everyone Needs to Know.* Oxford: Oxford University Press (2014), 114*ff* and George Lucas. *Ethics and Cyber Warfare: The Quest for Responsible Security in the Age of Digital Warfare.* (Oxford: Oxford University Press 2017), *passim*.

[9] Which it did, about which more later.

[10] Peter W. Singer and Allan Friedman. *Cybersecurity and Cyberwar: What Everyone Needs to Know.* (Oxford: Oxford University Press 2014), 119.

[11] Although this would have greatly increased the probability of its detection. In fact, this is how it was detected: it was found on computers other than those used in the Iranian nuclear enrichment program. (Peter W. Singer and Allan Friedman. *Cybersecurity and Cyberwar: What Everyone Needs to Know.* (Oxford: Oxford University Press 2014), 114*ff*.

[12] Cf. for example Ryan Jenkins, "Cyberwarfare as Ideal War." In Fritz Allhoff, Adam Henschke, and Bradley J. Strawser (eds.). *Binary Bullets: The Ethics of Cyberwarfare.* (Oxford: Oxford University Press, 2016).

command and control network, disrupting military communications but leaving all civilian infrastructure and communications systems intact. Or imagine a virus that exclusively attacks an enemy's targeting systems, such that all their guided munitions either explode harmlessly in the air or are re-directed to unpopulated areas where they do minimal damage. Such a weapon would seem to be a considerable ethical improvement in terms of respecting non-combatant immunity. Indeed, such a weapon might even make it possible to win a war while killing considerably fewer enemy combatants: not usually a concern of the Just War Tradition, but surely a consummation devoutly to be wished.

But we ought not begin firing up our hard-drives too quickly. Even though cyber weapons offer the *possibility* of extremely precise targeting, there are dangers here. First, note that Stuxnet presents another version of the Threshold problem discussed above. As it turns out, the Iranians did not respond kinetically to this attack and there were no human casualties on either side. But they might very well have done so. A shooting war might have started, with all the death and destruction attendant thereon. But second, even though cyber weapons can be incredibly precise, there is no guarantee that they will be so. Stuxnet was apparently an extremely resource-intensive weapon: it made use of at least four zero-day vulnerabilities, for example.[13] Zero-days are both very hard to find and (therefore) expensive (there is a well-developed black market for zero-days, with vulnerabilities in common operating systems reportedly going for upwards of a quarter of a million dollars). Other methods of "hacking" into enemy systems—determining MAC addresses or passwords, for example—can be equally resource-intensive, often requiring sophisticated human intelligence (HUMINT) methods.[14] An adversary may not be able to afford, or may decide not to invest the resources necessary, to procure and exploit the vulnerabilities necessary for a truly sophisticated—that is to say, precise—cyber attack. Arguably, every nation (or better, every party engaged in an armed conflict) ought to employ the most discriminate

---

13 Peter W. Singer and Allan Friedman. *Cybersecurity and Cyberwar: What Everyone Needs to Know*. (Oxford: Oxford University Press 2014), 115. A zero-day vulnerability is a vulnerability in software that is unknown to the software manufacturer—zero days have elapsed since the manufacturer became aware of the vulnerability—and so the manufacturer has not yet had a chance to "patch" the problem.

14 This is not to deny that such intelligence can sometimes be acquired quite cheaply. Phishing expeditions, for example, cost next to nothing. But sophisticated spear-phishing attacks, say, may require considerably more investment of resources, and it seems a safe generalization to suppose that the more crucial the vulnerability, the more difficult it will be to acquire.

weapons available—and no party may ethically or legally deploy weapons that are inherently indiscriminate—but it does not obviously follow from this that every nation—much less every non-state actor—ought to invest considerable resources into the development of humane weapons. And it is simply no use railing against a nation that they ought to have exclusively used laser-guided munitions, say, if they just don't have any in their arsenal.

Moreover, even if a nation does invest considerable resources into developing sophisticated and extremely precise weapons, there are no guarantees here. Modern code is—to use the technical term—"buggy." Any software program of any complexity will consist of literally millions of lines of code, and even a modest increase in the length of a program will usually result in a much greater increase in complexity, as every line of code potentially interacts with all the others. This means that there is considerable risk of unintended consequences when using cyber weapons. The best analogy here might be with biological weapons. While it might be possible to "tailor" a bio-weapon so that it only targets a specific group of people, given the enormous complexity of the genetic code and of environmental interactions, we should be extremely hesitant to even try to do so (assuming such a weapon would not be objectionable on other ethical grounds—which surely it would be). This is not to deny that cyber weapons can be targeted precisely, but even so sophisticated a weapon as Stuxnet apparently managed to "slip its lead" and infect computers across the world.[15] Thus, caution and due diligence, rather than unfettered optimism, are surely required here. It has apparently even been possible to reverse-engineer some features of Stuxnet. This introduces a new complication. Unlike conventional munitions, cyber weapons will often survive their use. This means that there are real worries about proliferation: if a cyber weapon falls into the wrong hands, there is considerable potential for it to be used for criminal, as well as legitimate military, purposes. Thus, the *in Bello* proportionality calculation gets a lot harder too. If we think in terms of conventional weapons, it's usually fairly clear what the consequences of use are. Most munitions have a fairly well-defined blast radius, allowing planners to calculate their likely effects. Even with nuclear weapons, we have a tolerably good understanding of the production and consequences of radiation. But with cyber, especially given the massive

---

15 "Stuxnet was specifically tailored to target just a few Iranian centrifuges and yet ended up spreading to well over 25,000 other computers around the world" (Peter W. Singer and Allan Friedman. *Cybersecurity and Cyberwar: What Everyone Needs to Know.* (Oxford: Oxford University Press 2014), 132.

interconnectedness of the internet, there will likely always be a greater element of unpredictability.

In any case, not all uses of cyber, even as part of conventional military operations, are likely to be as "clean" as Stuxnet, or even as the Operation Orchard attack. In 2008, when Russia attacked Georgia, the invasion was preceded for weeks by a series of cyber attacks against Georgian governmental infrastructure. It may be objected that many of these attacks were illegitimate, in that they were directed at civilian targets, but really what the case shows is that the majority of cyber targets are in fact "dual use," serving both military and civilian ends. This means that discrimination may actually be more difficult in the cyber realm in many ways. Even if we disallow direct attacks on "purely" civilian targets, such as financial institutions, the lines will often be difficult to draw. Would we permit, for example, an attack on a military pay system? On the one hand, military systems are generally considered fair game. On the other, preventing soldiers from getting paid seems to hit them quite "close to home," given that it will probably impact their families more directly than it impacts them: militaries generally feed, clothe, and house personnel who are engaged in an armed conflict, but this is often not the case for their families. An attack that disables a power grid, whether through kinetic or cyber means, can confer a considerable military advantage. But it can obviously also be extremely disruptive, even dangerous, to civilian life. The legal status of such attacks is somewhat in dispute, but states have argued for and engaged in them—witness the U.S. disabling of Iraqi power stations in the 2003 U.S.-led invasion—and it seems unduly optimistic to suppose that states will forswear their use in the future.[16] Rather, they will focus on the military advantage to be gained and discount the costs to noncombatants, especially those in the longer term. From an ethical perspective, however, the degree of human suffering such attacks cause must be taken into account. Without electrical power, hospitals can't operate, food spoils, water purification systems fail, etc. If this goes on long enough, people die.

GPS is a nice case in point here. The Global Positioning System—a series of satellites that allows precise location of points almost anywhere on the earth—was originally developed by and for the military. Disabling the

---

[16] See Kenneth R. Rizer, "Bombing Dual-Use Targets: Legal, Ethical, and Doctrinal Perspectives", Air University, 2001 for a nice discussion of the history and the issues here. Available at https://www.airuniversity.af.edu/Portals/10/ASPJ/journals/Chronicles/Rizer.pdf.

system, or at least rendering it unavailable to one's enemy, would provide a substantial military advantage, so long as one had some compensatory system available, either an alternate system or perhaps by retaining one's own access to the system while denying it to the enemy. The adversary would be unable to target its weapons precisely, locate its own forces with precision, etc. But the disruption to the relevant civilian sectors could be enormous. It's not just that Waze might not be available to guide one across town for discounted beer. Ships rely on GPS to navigate tricky channels, farmers rely on it to plant their fields, and it is integral to air traffic control systems worldwide. Without it, planes might crash and supply chains would certainly be disrupted. It is estimated that,

> A loss of satellite navigation for five days would cost the UK alone more than £5.1bn ($6.5bn), according to an assessment by the London School of Economics for the British Government. A failure of the GPS system would also cost the US economy an estimated $1bn (£760m) a day, and up to $1.5bn (£1.1bn) a day if it occurred during planting season for farmers in April and May.[17]

The point is that it can be very difficult to separate military and civilian uses of a (cyber) technology, and many military cyber targets are likely to involve profound civilian consequences. This makes both discrimination and proportionality problematic: both are likely to be more complicated and difficult to respect in the cyber realm, and it seems likely that there will be a tendency to implicate civilians more and more in armed conflicts.

The line between legitimate and illegitimate targets threatens to be blurred from the other direction, as it were, as well. Just as there will likely be an increased temptation to target civilians who are not legitimate targets (albeit indirectly, by targeting their critical infrastructure), there will almost certainly be civilians[18] who are legitimate targets, and this will also make discrimination more difficult. While many modern militaries have their own cyber components—the U.S. Army has its Cyber Command, the U.S.

---

17 David Hambling, "What Would the World do Without GPS?", BBC, October 2020, https://www.bbc.com/future/article/20201002-would-the-world-cope-without-gps-satellite-navigation.
18 I use "civilians" here, rather than the arguably more precise term "noncombatants" to make a point. It is of course not new that some civilians—for example, people not part of the military and not in uniform—are combatants. But whenever non-uniformed combatants participate in hostilities, lines get blurred. One of the primary purposes of uniforms, after all, is to mark one out as (almost surely—the notable exceptions being members of the clergy and medical personnel) a legitimate target. The increased use of cyber promises greatly to expand the number of such non-uniformed—"civilian"—combatants.

Navy has Tenth Fleet, China has PLA Unit 61398, etc.—the discipline and constraints of a military life do not appeal to all, and there are some very smart people—some people very good at cyber—who have no desire to join the military. This confronts States with a choice: unless they are willing to conscript such people, they will either miss out on a considerable talent pool, or they will have to find some other way of utilizing that talent. Many, perhaps most, State-level cyber operations today are carried out by non-military personnel,[19] and it seems naïve to think that in any serious conventional conflict, militaries would restrict themselves to the talent within their ranks. Thus, we are likely to see an increased use of personnel who are directly engaged in (the cyber component of) warfighting, but who are not themselves members of the military. And if such personnel are directly involved in and/or instrumental in enabling a conventional attack, one could hardly expect the opposing military to forgo attacking them.[20] Even if we confer combatant status on such individuals and consider them legitimate targets, they will often live and work as civilians: they may well not work on a military base, they will probably live at home with their families, etc. This will make it difficult to target them precisely, which in turn puts increased pressure on the principle of discrimination. If a State relies on combatants scattered, as it were, among its civilian population, it loses some standing in complaining about attacks that harm that population. If such targets are (legitimately) considered "high-value" enough, this may change the proportionality calculation to the point where even relatively indiscriminate attacks on them can be justified.

Thus far, we have been thinking primarily in terms of existing technology and uses of cyber, but it may be worth trying to look down the road a bit; there is no doubt that cyber will continue to develop, probably at an increasing rate. As weapons systems become more and more computerized, they will become less and less under direct human control and in this sense at least more and more autonomous. Indeed, it seems likely that relatively soon we will see the advent of fully autonomous weapons systems (AWS) on the battlefield—systems that are not under direct human control but can operate on their own (within pre-specified parameters) and make their own decisions, including decisions about whether and whom to kill. Much

---

19 Sometimes, of course, even officially non-governmental personnel. Some States prefer to utilize such actors for the "plausible deniability" their use affords.
20 Assuming that military can determine who they are. To the extent that it can't, this is potentially a further temptation toward indiscriminate—or at least less-than-fully discriminate—behavior.

has been written about the morality of AWS, and I will not repeat (much of) it here. While critics of AWS frequently point out that autonomous systems will never be perfect and will therefore at least sometimes make "moral" mistakes, as Ron Arkin points out,[21] it is entirely plausible that they will nevertheless be able to outperform human beings. Robots never get tired, scared, bored, angry, or vengeful, and even autonomous "self-driving" cars, for all that their occasional accidents make headlines, generally have a safety record that most humans would envy, having caused very few fatalities for millions of miles driven. Nevertheless, some people are absolutely horrified by the idea of machines being allowed to kill human beings entirely of their own accord—they see therein a massive violation of and an affront to human dignity—while others—this author included—are largely unmoved by the worry. I simply see no real moral, and certainly no practical, difference in being killed by an AWS or from a bomb dropped, albeit by a human, from 30,000 ft. However that may be, machines deciding whether or not to kill people is likely to become the case *de facto*, if not *de jure*. First, humans have a tendency to rely perhaps overly much on technology—witness the "instrument bias" that may have played a role in the United States' accidental downing of an Iranian passenger airliner in 1988; human operators apparently overrode their own misgivings and trusted the Aegis targeting system instead.[22] This sort of thing is likely to occur more frequently, given the "gee whiz" factor of increasingly sophisticated technologies. Such systems can seem so powerful and so impressive, that people may default to trusting them over their own better judgment. But second, and more fundamentally, cyber activities occur at clock speed, far faster than the processing capabilities of mere humans. In any conflict involving a robust cyber component, decisions will be made by computers in milliseconds. We live in an age of Multiple Independently Targetable Re-entry Vehicles (MIRVs) and (soon) hypersonic missiles. Any human or group of humans, trying to cut through the fog of war in a sizable conflict, will likely be overwhelmed. Multiple threats—both cyber and kinetic—will have to be identified, multiple targeting decisions will have to be made, seemingly all at once, else the conflict might apparently be over almost before it begins. Thus, computers will simply have to make some of the relevant decisions. And some of those decisions are quite likely to

---

21 Ronald C. Arkin, "The Case for Ethical Autonomy in Unmanned Systems." *Journal of Military Ethics* 9(4), 2010.
22 See, for example, "Iran Air Flight 655", Wikipedia, https://en.wikipedia.org/wiki/Iran_Air_Flight_655.

eventuate in humans being killed. In this sense at least, computers deciding whether and whom to kill is likely to become the norm.[23]

Jus post Bellum

Here the news would seem to be mostly good for a change. Bombs and bullets cause considerable and long-lasting damage, but they become no more destructive (and may even be more precise) when computers are involved. So there does not seem to be any reason to think that cyber capabilities will make warfare any more destructive (modulo the above worries) in the kinetic realm. And "cyber damage"—the damage done by computers to other computers and information systems—may be much less destructive and long-lasting than kinetic damage. Many cyber effects are completely reversible: systems can be patched, data can be restored, access can be re-authorized, etc. In sum, the *status quo ante* can be regained. Admittedly, not all cyber damage—again, damage done by computers to other computers—is reversible. Some data may be irretrievably lost; some systems may be irreparably damaged. But in the main, cyber damage does not consist in smoking craters and torn bodies. To the extent, then, that the establishment of a just and lasting peace requires reconstructing infrastructure and repairing damage, the cyber realm would seem to be a much more congenial—and indeed, a morally preferable—realm in which to operate.

Conclusion

As Niels Bohr once remarked, prediction is very difficult, especially if it's about the future. But barring some worldwide catastrophe—some return to the stone age or at least to an age of vacuum tubes and transistors—there seems almost no chance that the cyber genie will be put back in the

---

23 As an analogy, think of the role played by computerized trading systems in stock markets worldwide. Computers can aggregate and analyze vast amounts of data and make trades literally in milliseconds. Any human trying to compete "by hand," as it were, would be almost guaranteed to lose and to lose big.

bottle. Indeed, if Moore's law continues to hold,[24] the pace of cyber seems likely only to quicken, with cyber occupying a larger and larger part of our lives, and of warfare. And in warfare, as in life, this is to be both welcomed and feared. There is considerable potential for cyber to make warfare and the battlefield a more humane place. If more and more conflict happens "online," as it were, perhaps there will be less killing and physical destruction. But there is also potential for cyber to occasion widespread suffering and misery, if it proliferates the number of wars that are fought or expands the scope of those who are caught up in those wars. The underlying ethical principles and considerations that govern conflict and violence, it seems to me, don't really change when cyber enters the picture, but we will have to make sure that we pay attention to and apply them at the same rate at which the world changes.

References

Arkin, R. C. "The Case for Ethical Autonomy in Unmanned Systems." *Journal of Military Ethics* 9(4), (2010).

BBC, Hambling, D. "What Would the World do Without GPS?", October 2020, https://www.bbc.com/future/article/20201002-would-the-world-cope-without-gps-satellite-navigation.

Historynet, "Birds of War", Jan 2019, https://www.historynet.com/birds-of-war.htm.

Jenkins, R. "Cyberwarfare as Ideal War." *Binary Bullets: The Ethics of Cyberwarfare*, ed. Fritz Allhoff, Adam Henschke, and Bradley J. Strawser, (Oxford: Oxford University Press, 2016).

Lucas, G. *Ethics and Cyber Warfare: The Quest for Responsible Security in the Age of Digital Warfare*. (Oxford: Oxford University Press 2017).

Rizer, K.R. "Bombing Dual-Use Targets: Legal, Ethical, and Doctrinal Perspectives", Air University, 2001 for a nice discussion of the history and the issues here. Available at https://www.airuniversity.af.edu/Portals/10/ASPJ/journals/Chronicles/Rizer.pdf.

Schmitt, M.N. and L. Vihul. "The Emergence of International Legal Norms for Cyberconflict." *Binary Bullets: The Ethics of Cyberwarfare*, ed. Fritz Allhoff, Adam Henschke, and Bradley J. Strawser, (Oxford: Oxford University Press, 2016)

Singer, P.W. and A. Friedman. *Cybersecurity and Cyberwar: What Everyone Needs to Know*. (Oxford: Oxford University Press, 2014).

---

[24] Moore's Law is the observation, owing to Gordon Moore, one of the early computer pioneers, that the number of transistors on an integrated circuit (and thus, roughly, computer processing speeds) tends to double on average every two years. For all that it can't really be a law, the generalization has proved surprisingly robust over the last 50 or so years.

The New York Times, "As Understanding of Russian Hacking Grows, So Does Alarm", (Jan 2021), https://www.nytimes.com/2021/01/02/us/politics/russian-hacking-government.html.
Wikipedia, "Iran Air Flight 655", https://en.wikipedia.org/wiki/Iran_Air_Flight_655.
Wikipedia, "Iraq Inquiry", https://en.wikipedia.org/wiki/Iraq_Inquiry.

5

THE ETHICS OF CYBER-SABOTAGE

Jeremy Davis

Introduction

Cyber warfare has long been thought of as the future of war. But recent years have shown that the future has arrived. The next several decades are sure to witness an ever-increasing emphasis on cyber war as a strategy, both offensive and defensive, for nation-states and non-state actors alike.

Cyber warfare is different in many ways from ordinary, conventional warfare. Whereas wars throughout history have been fought on—and, in many cases, fought *over*—physical territory, global conflict is nowadays increasingly waged in cyber space, using local, national, and global information networks as the battlefield. Rather than targeting combatants or physical structures directly, cyber attacks generally target an adversary's technological infrastructure or certain of its key components: critical cyber networks are rendered inoperable, information systems are disrupted, and sensitive and important data is corrupted, destroyed, or intercepted.

These distinctive features of cyber warfare give rise to novel and pressing questions concerning long-term military strategy, training and technological development, and various other operational and tactical issues. But they also prompt a range of ethical questions as well. My focus in this chapter is a particular kind of cyber attack—namely, cyber-sabotage—and the particular issues it raises. By focusing on this particular tactic, we not only gain appreciation of the distinctive ethical questions it poses, but we also gain further insight into the broader ethical questions that arise elsewhere in cyber warfare.

What is "Sabotage?"

To understand the phenomenon of cyber-sabotage, we must first consider sabotage in more ordinary contexts. Here are some rather ordinary cases of sabotage:

1. An athlete or team pulls the fire alarm at the opponent's hotel in the middle of the night before the game, thereby disrupting the team's sleep and affecting their performance during the game (as the New England Patriots are alleged to have done on several occasions).
2. A student pre-emptively checks out all the relevant books from the library (which, according to some rumors, is typical at many elite law schools) so as to lower a rival classmate's chances for succeeding on a project.
3. A business owner writes scathing online reviews of a rival business anonymously (as is apparently common in the restaurant industry) in order to draw business away from them and toward her business instead.

What do these three examples have in common? First, they all identify cases in which someone (either an individual or a group) *deliberately attempts to frustrate the efforts of their adversary in order to gain a strategic advantage over them in competitive environments.* Notice that this is not done by directing efforts at the adversaries themselves. For example, we probably would not describe it as "sabotage" when Tonya Harding's associates beat her opponent, Nancy Kerrigan, with a lead pipe in an effort to prevent her from competing.

For an act to qualify as sabotage, it seems that one's efforts must be focused not on the adversary directly, but rather on the materials, environment, or circumstances necessary for them to carry out their plans. Our previous examples—disrupting an opponent's sleep, pre-emptively checking out library books, and leaving fake reviews—all possess this feature.

Furthermore, despite the possibility of being suspected or even identified as the culprit, those carrying out the sabotage—the saboteurs—in each of these cases *are either undetectable, have plausible deniability, or benefit from some other method for avoiding detection.* In some cases, being identified as the saboteur will render the attempted sabotage ineffective or will generate justifications for retaliation on the part of the victim of the sabotage. In many cases, it might be obvious to the victim who the saboteur was, even if it cannot be proven decisively—which may be enough to qualify as a success on the part of the saboteur, depending on the circumstances. In any case, the efforts to avoid detection are central to sabotage efforts.

This all leads us to the following rough definition of sabotage:

> **Sabotage:** an act that aims to weaken, damage, destroy, or otherwise frustrate the use of equipment, systems, environment, or other conditions necessary or useful for success by an adversary, so as to gain a strategic advantage over the adversary in a competitive environment, with an effort to avoid detection or attribution.

That's a bit too wordy and complex; so, at the risk of losing some of the nuance, let's put it more simply:

> an act of sabotage is an attempt to prevent your adversary from succeeding against you by messing with their materials or environment without them knowing that you were the one who did it.

As our earlier examples show, sabotage is a common feature of many different competitive environments—from sports to business to interpersonal rivalries. But sabotage is perhaps best well known as a military tactic. State and non-state actors alike attempt various acts of sabotage during the course of a conflict, prior to the onset of a conflict, or even in peacetime. Acts of sabotage between global adversaries throughout history include such things as bombing munitions factories; setting buildings on fire; sinking ships; disrupting or corrupting oil, water, or other resource systems; damaging or destroying railcars and railways; interfering in communication networks, such as radio signals; and countless other acts like these.

Note that, unlike the cases of sabotage from everyday life, it is not uncommon for military sabotage to lead to quite significant harms, including deaths. For example, it would be sabotage to disrupt the fuel line on an adversary's military aircraft. And this might later result in the death of an enemy pilot, if he takes to the skies without noticing that he's been sabotaged. Indeed, this might be precisely the desired outcome of the sabotage attempt. But having the intention to produce this sort of effect does not mean the initial act ceases to be one of sabotage. The target, after all, was still the aircraft itself, not the pilot. This might, however, create a shift in our moral assessment of the attempt—but we'll discuss that in more detail later on.

What is "Cyber-Sabotage"?

What, then, is *cyber*-sabotage? Given that the introduction of cyber technologies in military contexts is a relatively new phenomenon, terms like "cyber war" and "cyber-sabotage" tend to be used in different ways by different people, and the parameters of these concepts are debated among scholars and practitioners.

One way of trying to define cyber-sabotage is to ask: What does it mean to attach "cyber-" to a concept? For example, assuming we know what an act of war is, what is suggested when we call something an act of "cyber-war?" Two particular features spring to mind.

First, the *means:* to say that something is an act of "cyber-war," "cyber-sabotage," etc. suggests that one uses cyber means to carry out that act. That is, whereas ordinary acts of sabotage can use a variety of means to frustrate an adversary's efforts, to engage in *cyber*-sabotage is to use cyber technologies—such as the internet, information systems, or other such technologies—to carry out the act.

Second, the *target:* cyber wars generally have consequences for the physical world, but their initial or primary target is located in the cyber realm. So, while cyber wars can cause real physical destruction, this is made possible by attacks that occur to or within cyber systems, networks, and other related technologies. Thus, an act of cyber-sabotage is one in which the target is elements of an adversary's cyber systems, like an adversary's information or communication networks, the underlying cyber infrastructure of various physical processes or systems, and so on.

Generally, "cyber-" acts involve *both* of these features. For one thing, it is difficult to imagine many cases that use cyber means without a cyber target. But there are certainly cases of acts that have a cyber *target*, but do not use cyber *means*—for example, purely physical attacks an adversary's information servers. Cases like these are just more ordinary instances of sabotage, rather than *cyber-sabotage*.

This all leads us to the following definition:

> **Cyber-sabotage:** an act of sabotage (as previously defined) in which both the *target* of the act and the *means* of carrying out the act center on cyber technologies, digital infrastructure, online information systems, or other related systems.

Now that we have a rough definition in hand, let us consider two recent high-profile cases of cyber-sabotage—namely, *Stuxnet* and *Operation Orchard*.

Case Studies: Stuxnet and Operation Orchard

Stuxnet

In 2010, a Belorussian cyber security company discovered a unique and sophisticated piece of malware, which came to be known as *Stuxnet*. The users who first noticed the virus, however, were not the intended targets of the virus. Indeed, they were the unintended collateral damage of a different target—namely, the systems at the heart of the Iranian nuclear program.

In fact, Stuxnet had a very specific target: the programmable logic controllers (PLCs) that monitor and regulate the machinery used at Iran's nuclear plants. These controllers are responsible for the speed of the uranium centrifuges that facilitate uranium enrichment, which are a key step in the production of nuclear weapons.

Once it gained access to the network—a piece of the puzzle that is still not entirely known—Stuxnet moved through every computer on the network and determined whether or not a given system employs a certain sort of PLC. If it does, then the virus embeds itself into the code of that PLC, and quietly changes certain of its functions—most notably, the speed of the centrifuges. These changes initiate a process where the centrifuge speeds up and then slows down over and over, which significantly disrupts the uranium enrichment process and gradually damages the centrifuges themselves. Finally, Stuxnet ensures these changes do not appear on the system's reports, thereby ensuring that this whole process goes undetected.

In terms of its immediate goals, Stuxnet appears to have been a success. Many of the centrifuges at Iran's nuclear facility in Natanz were significantly damaged, which helped to delay Iran's nuclear program. This delay, however, was only temporary: Iran's enrichment efforts were back to normal within a year.

Critical to the virus's success was the exploitation of several "zero-days"—the technical term for a vulnerability that is unknown to that system's user. Stuxnet exploited an unprecedented *four* zero-days in the Iranian system. For many cyber security experts, this is compelling evidence that the attack could not have been carried out by a rogue agent, or even a small group of hackers. Rather, it must have been carried out by a state entity, or perhaps multiple states working in concert. Many analysts believe Stuxnet was a joint product of both the American and Israeli governments, though neither party has explicitly claimed responsibility.

Many have called Stuxnet one of the world's first major cyber-attacks; some think it could qualify as an act of war. But it is also a very clear example of cyber-sabotage. After all, the apparent goal of Stuxnet was to disrupt, in a surreptitious manner, Iran's nuclear enrichment so as to gain an advantage over them—in this case, by delaying their ability to strengthen their nuclear capabilities. Moreover, it was carried out using cyber means, and targeted the cyber networks that regulate the operations of their centrifuges. As Michael Joseph Gross put it: "Stuxnet is like a self-directed stealth drone: the first known virus that, released into the wild, can seek out a specific target, sabotage it, and hide both its existence and its effects until after the damage is done. This is revolutionary."[1]

Operation Orchard

As we just saw, Stuxnet is an especially clear example of cyber-sabotage. But whereas Stuxnet occurred during peacetime, there are also instances in recent history of cyber-sabotage that was a part of direct military action. One such case occurred in 2007, as part of an Israeli airstrike on a Syrian nuclear reactor, in a secret operation known as "Operation Orchard" (or "Operation Outside the Box").[2]

Israel has long had concerns about Syria's nuclear ambitions. In 2006, Israeli intelligence became aware of an apparent nuclear reactor in Al Kibar, Syria. Nearly a year later, Israeli Special Forces conducted a covert operation in which they infiltrated the facility and collected water and soil samples, which was then tested for chemical evidence to determine the status of the development of the reactor. These samples showed that the reactor was not yet fully operational, though it was likely to be soon. Recognizing the potentially small window of opportunity, the Mossad determined that an airstrike on the facility was the best option for preventing Syria from obtaining nuclear capabilities.

---

1 Gross, Michael Joseph. "A declaration of cyber-war." *Vanity Fair* 53, no. 4 (2011).
2 For more details, see: Follath, E. and H. Stark. 2009. "The Story of 'Operation Orchard': How Israel Destroyed Syria's Al Kibar Nuclear Reactor." Spiegel Online International, November 2. http://www.spiegel.de/international/world/the-story-of-operation-orchard-how-israel-destroyedsyria-s-al-kibar-nuclear-reactor-a-658663.html.

It is at this point that the element of cyber-sabotage comes into the picture. Unbeknownst to the Syrians, the microprocessors that they used in their radar technology contained a hidden kill-switch, which had been pre-programmed into it during its development. Activating the kill-switch allows one to generate a false radar picture, which can be used to avoid detection when entering the user's airspace.

The kill-switch was unknown to the Syrians, but the Israelis certainly knew of it—and more importantly, knew how to exploit it. Prior to the airstrike, the Mossad activated the kill-switch, which allowed the Air Force to enter Syrian airspace, carry out the mission, and exit—all without being detected.

From the Israeli perspective, the airstrike was a success: the facility at Al Kibar was completely destroyed. There were, however, unconfirmed reports of between 10 and 24 North Korean nuclear scientists who were unintentionally killed during the bombing.

So, while the centerpiece of Operation Orchard was a conventional airstrike, it was made possible by the multi-stage cyber-sabotage that Israeli intelligence employed.

What is Distinctive about Cyber-Sabotage?

With these examples in hand, we can now explore some of the features that make cyber-sabotage distinctive, and thus, a particularly attractive tool for contemporary warfighting.

First, cyber-sabotage often allows for an advanced level of *undetectability*. It is of course essential to most sabotage attempts that one avoids detection before, during, and after the attempt. If the saboteur is detected at any of these stages, the adversary might have time to thwart or prevent the attempt, or it could significantly weaken the effectiveness of the attempt.

Of course, artful saboteurs can avoid detection in physical cases of sabotage. But the cyber realm may afford an even greater opportunity for avoiding detection in many cases. Cyber-sabotage often centers on exploiting vulnerabilities in the adversary's software or systems that are either unknown to them or are otherwise difficult to keep secure. Cyber-saboteurs are therefore able to access and disrupt those with decreased likelihood of detection. And if the saboteur is indeed able to exploit the vulnerability without detection, they are likely to have the means to "leave no trace" as well.

Stuxnet and Operation Orchard are both great examples of the element of undetectability. The unprecedented four zero-days allowed the cyber-saboteurs to infiltrate the Iranian networks while avoiding detection. The use of the kill-switch in Operation Orchard had a similar effect for Israel.

Furthermore, since their physical presence behind enemy lines is not typically required, cyber-saboteurs are at much lower risk of being captured or harmed in the course of the attempt. In cases that do not go according to plan, or that require some physical-world presence, cyber-saboteurs may be detected. For example, it is assumed that the Stuxnet entered the Iranian network via a USB drive, which must have been carried out by someone on the ground at the facility. And yet, when compared to physical world sabotage attempts, the risks of harms upon being caught to any of those part of a cyber-sabotage mission are much lower. Adversaries are generally afforded fewer opportunities to detain, threaten, injure, extract information from, or otherwise use the cyber-saboteur for their own ends, as compared with other physical tactics.

This leads us to another distinctive feature of many cases of cyber-sabotage: *deniability*. For many of the same reasons that cyber-saboteurs more easily avoid detection, it is also easier for them to avoid having their acts of sabotage attributed to them. The best cyber-saboteurs typically leave behind very little or no evidence that can be directly traced back to their state. There are rarely any digital "fingerprints."

In some cases, deniability will not be a relevant goal of the cyber-saboteur: Operation Orchard is a clear instance in which it was apparent who was behind the attack, so there would be little chance of Israel denying using the kill-switch. And in many other cases, the victims of sabotage will have their suspicions. It is clear to most observers, including Iran, that the U.S. and Israel must have been behind the Stuxnet attack, though neither party has admitted their role. In many cases, it may be relatively obvious who is behind the attack, based on the strategic value of the target, the specific tactics employed, and so on. Nevertheless, states will often be able to deny their involvement. They may even plausibly claim it was a rogue actor, or some other unaffiliated non-state entity.

Deniability is especially valuable when cyber-sabotage is preventive in nature—that is, when it takes place outside of the context of an ongoing, officially declared war. This allows states to take action to prevent or diminish emerging threats from adversaries, such as nuclear weapons development, without being accused of having performed an act of war. This significantly limits what that adversary is able to do—legally or morally—in response.

Of course, this is not always guaranteed. Indeed, it is thought that Iran engaged in retaliatory cyber-attacks on American banks in the wake of the Stuxnet attack, though the exact relationship is uncertain. At any rate, when compared with physical sabotage, or other related physical attacks, cyber-sabotage generally affords those who employ it a greater deniability, and in turn, a decreased likelihood of significant retaliation.

One final distinctive element of cyber-sabotage is its *scale*. Most of the central targets of cyber-sabotage are an adversary's information systems or other online networks, which are typically heavily integrated across institutions and society as a whole. It is possible, then, that even those acts that may seem like small disruptions can wreak extraordinary havoc. In other words, the scale of the effects of many acts of cyber-sabotage can be enormous. Once again, Stuxnet provides a great example of this: the havoc one computer virus can wreak can be substantial and can devastate an entire nuclear facility for quite some time.

To be sure, it is not impossible, in cases of physical sabotage, to generate significant effects from relatively small acts. What may seem like minor disruptions to physical systems, such as the water supply, can generate enormous problems for an adversary. But, again, the scale of the effects possible in cyber-sabotage can easily dwarf that of physical sabotage. There is a tremendous difference in the scale of a mostly local disruption in the power grid or water supply, and the more national or even global disruption that is possible through cyber-sabotage. This is not to say that every act of cyber-sabotage aims to be maximally disruptive in this way. The point, however, is that the possibilities with cyber-sabotage are much broader.

To sum up: cyber-sabotage has at least three features that make it especially appealing—namely, undetectability, deniability, and scale. This does not mean that all three of these are present in every case of cyber-sabotage. Nor does it mean that there are no other features of cyber-sabotage that may be thought to be appealing. What it does mean, however, is that cyber-sabotage is a unique tool in the military arsenal for fighting wars in the modern era.

Ethical Questions

So, cyber-sabotage offers a number of advantages, but it also raises several distinctive ethical issues. In this section, we will explore three questions concerning the ethics of cyber-sabotage:

1. How well do the central conditions of just war theory capture cyber-sabotage?
2. In light of some of the features of cyber-sabotage discussed above, how should we understand moral responsibility for the harms of cyber-sabotage?
3. What ethical concerns arise from the prospect of cyber-sabotage becoming a more commonly used military strategy in the future?

Before we dive in, it is worth noting that our discussion will center on *ethical* or *moral* concerns—not *legal* ones. To be sure, there are a number of questions concerning cyber-sabotage's status in international law, but we won't be looking at those issues directly here. For one thing, the legal understanding of cyber sabotage is currently undergoing rapid evolution: anything written now on the legality of cyber sabotage will surely be outdated within a short time. Furthermore, many philosophers, ethicists, and lawyers believe that our laws should reflect, or at least be informed by, ethical theory. Thus, to get at the legal questions, it makes sense to start with the ethics.

1. How well do the central conditions of 'just war theory' capture cyber-sabotage?

The most influential framework for thinking about ethics in war is called "just war theory." It has its roots in theologians like St. Augustine and St. Thomas Aquinas, and has served as the basis for much of the law of war, as well as most contemporary discussions of the ethics of war. This theory offers a set of conditions that constrain permissible entry into war, which are known as the *jus ad bellum* conditions, as well as constraints on permissible conduct during war, which are known as the *jus in bello* conditions.

Contemporary writers in the just war tradition disagree somewhat on which of these conditions apply, but the standard articulation of these principles is as follows:

Jus ad bellum:

1. **Just Cause:** a war can only be undertaken for particularly good moral reasons
2. **Proportionality:** the harms that the war will do must not be out of proportion to the good it will achieve

3. **Necessity/Last Resort:** the war must be the option of last resort, which means all other less harmful alternatives must have been explored and exhausted prior to entering into war
4. **Reasonable Chance of Success:** wars must not be undertaken if futile; they must stand a reasonable chance of success at achieving at least some of the just cause(s)
5. **Legitimate Authority:** wars cannot be declared by just anyone; they must be declared and carried out by the relevant authority
6. **Right Intention:** wars must not be undertaken with immoral or irrelevant intentions.

Jus in bello:

1. **Proportionality:** just like the *jus ad bellum* version of this condition, but applied to particular acts in war (as opposed to the war on the whole)
2. **Necessity/Last Resort:** just like the *jus ad bellum* version of this condition, but applied to particular acts in war (as opposed to the decision to enter the war)
3. **Distinction/Discrimination:** civilians enjoy privileged status in war, and ought not to be targeted directly. Moreover, harms to civilians ought to be minimized, even at cost to soldiers themselves.

While these conditions predate the cyber-era by hundreds of years, they are timeless in their application. Even acts done with novel technologies, such as cyber-sabotage, must satisfy these criteria to be morally permissible.

How would just war theory judge the permissibility of cyber-sabotage? Let's look first at the *jus ad bellum* criteria, focusing particularly on the *just cause* condition. Stuxnet seems to offer a helpful case study here: there was no ongoing war at the time of the attack, which means that the attack itself can be understood as constituting the initial strike, initiating hostilities between two previously peaceful nations. (To simplify, let's assume for the moment that the U.S. and Israel were behind the attack—though, again, they both deny their involvement.)

Was Stuxnet an unjust attack? This can be answered by referring to the just cause condition. Typically, this condition allows states to interfere with the sovereignty of another state (or use defensive force against its officials, like soldiers) only in order to (1) prevent or stop an ongoing or imminent unjust attack against oneself or a third-party, or (2) to protect against ongoing or imminent widespread human rights abuses.

But Stuxnet was neither of these. Apparently, it was intended only as a way of slowing the development of Iran's nuclear program. While this may be an important step in protecting against *future* attacks, the attack did not prevent an *imminent* threat of one of the aforementioned kinds. Thus, if it *was* an act of war, there is a case to be made that it failed to satisfy just cause.

Does this mean that Iran had just cause to retaliate against the U.S. and Israel? The answer to this question is complicated and will depend significantly on how one understands the limits of these just war conditions.

First, notice that one distinctive feature of most cyber-attacks—particularly acts of cyber-sabotage—is that they are not typically *ongoing*. Once the victim has become aware of the attack, it is already over. On one view, for Iran to respond by launching a strike against the U.S. or Israel in retaliation for this attack (as they are alleged to have done) *also* violates the just cause condition, since there is no ongoing threat of harm to prevent. Doing so therefore amounts to an act of reprisal, which is not consistent with the requirements of the just cause condition as it is typically understood.

An alternative view holds that such retaliation could indeed qualify as just. One view is that it constitutes punishment for injustice. In domestic legal contexts, we typically think punishing wrongdoers after they've committed wrongful acts is acceptable, even when they are not immediately threatening further harms. Many contemporary writers in the just war tradition, however, do not believe that punitive wars are just. This condition is, on this view, "forward-looking": states must aim to prevent future (or at least ongoing) threats, not rectify past wrongs.

One could argue instead that retaliation constitutes defense against continuing hostilities, or deterrence against future attacks of this sort. Such rationales make sense if there is reason to view the initial attack as the initiation of a pattern of hostilities. This is surely how we understand certain kinetic attacks, such as the attack on Pearl Harbor by the Japanese. Few would have argued that the Allies were prohibited from responding; indeed, the attack itself clearly constituted the beginning of hostilities. But it is unclear whether this same rationale is available to Iran in the Stuxnet case. It is difficult to know whether the attack was isolated or one of many more to soon follow. This difficulty is exacerbated by the fact that, again, no one took responsibility for the attack.

These points underscore one of the central ethical challenges for cyber-attacks and cyber-sabotage in particular: in addition to there being significant practical challenges in preventing such attacks, there are also ethical challenges in responding to them. As we saw, there are arguments on both sides, but the ethical boundaries are far from obvious. On the more restrictive view, cyber-attacks essentially preclude *any* sort of permissible defensive response by the victim state. This might be thought to be too restrictive. The less restrictive view, however, seems to open up the just cause to many other, much-less appealing rationales for war, such as punitive wars. There is therefore a significant ethical question that requires further thought and development.

One might instead argue that such attacks are better viewed as instances of "force-short-of-war." States use such acts all the time, including diplomatic pressure, sanctions, no-fly zones, and trade embargoes, among others. These are not military attacks or initiations of hostilities, but rather attempts at preventing full-scale war via less harmful alternatives. Cyber-sabotage, or cyber-attacks more generally, might be thought of as a somewhat more direct version of force-short-of-war.

Force-short-of-war is valuable in part because it is typically much less harmful than war. But it is also an important part of satisfying the aforementioned necessity (or last resort) condition. This condition, recall, holds that all other less harmful alternatives must have been explored and exhausted before resorting to war. Thus, options like diplomatic pressure, sanctions, and so on must be explored, if possible, before a state can permissibly start a war.

Stuxnet seems like a good example of force-short-of-war. Prior to the attack, Iran was undeterred from building its nuclear arsenal, despite repeated diplomatic efforts (including sanctions and trade embargoes) by Western governments like the U.S. Were Iran to develop a nuclear weapon, this could destabilize the region and pose a grave threat to the U.S. and others. Stuxnet was somewhat more destructive than those other diplomatic alternatives; but then again, Stuxnet was only attempted after those other options failed. Furthermore, as we saw, Stuxnet was incredibly limited in its damage, physical or otherwise, beyond the Natanz facility. Of course, it was still a somewhat aggressive act against the government of a sovereign nation. It is therefore a challenge how best to classify acts of cyber-sabotage like this one. Different ways of classifying it reveal different ethical dimensions of the action.

Some thinkers have sought to spell out an alternative framework or set of conditions—like the *jus ad bellum* and *jus in bello* we noted above—to accommodate cases of force-short-of-war. A theory of *jus ad vim*, as this is called, offers the criteria for just use of force of all kinds, rather than just those more narrowly included within the category of "war." Compared to the other central conditions of just war thinking, *jus ad vim* is in its infancy: its entry into the mainstream discussion came in 2006, in an updated preface to the classic *Just and Unjust Wars* by Michael Walzer.[3]

On certain versions of *jus ad vim*, the essence of the just war criteria are maintained, but adapted to fit a broader context in which many different encounters, lethal and otherwise, might occur (even if none rise to the level of "war").[4] A central feature of this revised approach is the emphasis on avoiding escalation: force-short-of-war is impermissible if engaging in it yields a high probability of resulting in war. Of course, whether one's adversary escalates in retaliation is hard to predict. Moreover, as we will see in a different context in the following section, it is unclear whether one's current actions can be made wrongful by the later independent actions of another agent.

There is much more that can be said about *jus ad vim*. Indeed, given its recent entrance into discussions on the ethics of war, combined with its relevance to a range of current events, scholars will surely have much more to say on this in the coming years. It will suffice here to note simply that the ethical questions surrounding Stuxnet, and other attacks like it, are much more complex than they may have initially seemed.

2. In light of some of the features of cyber-sabotage discussed above, how should we understand moral responsibility for the harms of cyber-sabotage?

Often when cyber-saboteurs are successful, their victims are not aware of the sabotage until it's too late. As we saw with Stuxnet, this meant the destruction of many important centrifuges. In Operation Orchard, it meant the Israelis were able to destroy a Syrian nuclear facility, killing North Korean scientists in the process. In many other cases, various harms—some intended, some merely foreseen side effects, and others unforeseen—will follow from acts of cyber-sabotage.

This prompts the question: Is the cyber-saboteur morally responsible for all the harms that stem from their act of cyber-sabotage? There are many issues to explore here, but let's focus on some of the most central.

---

[3] Michael Walzer, *Just and Unjust Wars*, fifth ed. (New York: Basic Books, 2015.
[4] A recent version of this view can be found in Daniel Brunstetter and Megan Braun, "From Jus ad Bellum to Jus ad Vim: Recalibrating Our Understanding of the Moral Use of Force," *Ethics and International Affairs* 27, no. 1 (2013), 87-106.

One might be tempted to claim that saboteurs are only morally responsible for the harms they directly intend. The saboteurs behind Stuxnet clearly intended that the centrifuges malfunction, and that is exactly what happened. But what if the sabotage had caused a catastrophic failure that led to many innocent deaths?

If we think that cyber-saboteurs are only responsible for deaths they intend, then they wouldn't be responsible for these deaths. But we don't typically think that individuals are morally responsible only for the harms they intend; rather, they are responsible for most (or, on some views, all) of the reasonably foreseeable harms that stem from their actions. In some cases, these harms are enough to render the initial action impermissible. Thus, cyber-saboteurs must surely be responsible for many more of the effects of their actions, even those they hadn't intended. In our hypothetical deadly Stuxnet scenario, the cyber-saboteurs would be responsible for the innocent deaths.

One complicating factor here is something that is common (though certainly not unique) to many cases of sabotage: some of the harmful effects are not simply natural consequences of one's actions, but rather are brought about through the agency of other people. Philosophers call this phenomenon "intervening agency." It is a common feature of our moral lives, and it occurs often in war or other conflict scenarios. Consider a simplified case: A attacks B, which prompts B to attack some third-party, C. Is A responsible for the harms to C?

Some people might be inclined to say, "Well, that's B's responsibility. A can't control what B does." This may be true in some contexts, but what about the context of cyber-sabotage? As we have seen, one of the distinctive elements of cyber-sabotage is that it often relies upon the victim remaining unaware that anything has gone awry and acting as though everything is normal. The problem, of course, is that it is hard to predict exactly how people will behave; whether or when they will take notice of the sabotage; and how they will respond to it if they do. It is not uncommon for acts of sabotage to involve effects that weren't foreseen in the planning stages. In short, there is enormous uncertainty as the effects of cyber-sabotage unfold.

Some of the harms that follow from acts of cyber-sabotage are the responsibility of the victims. This is clearest when the harms in question are responses to the attack, rather than more direct consequences of it. Taken too far, however, this point comes dangerously close to a problematic form of "blaming the victim." For example, it is implausible to think that states should be held responsible for the harms stemming from deliberately undetectable hacks on their financial networks. To be sure, states should take relevant precautions to avoid such occurrences; but there are certain things we cannot hold victims responsible for.

Again, this phenomenon is familiar in war more generally, though it takes on a new form in the case of cyber-sabotage. The upshot is that cyber-saboteurs cannot claim "clean hands" when their acts spiral out of control. Indeed, prior to undertaking such acts, prospective saboteurs must ensure that hey have accounted for the full range of potential harms that might ensue. This requires, at a minimum, the same sort of analysis that is required in kinetic attacks—namely, an inquiry into all foreseeable harms, casualties (where applicable), and other downstream effects.

One final point on moral responsibility: we have been speaking here about an objective form of moral responsibility—who is *in fact* responsible. But there's another notion of moral responsibility—namely, whether members of the moral community can hold one another morally responsible. The element of deniability in cyber-sabotage renders this particularly difficult. When we cannot know for certain who is behind an attack, it becomes nearly impossible to hold the attackers morally responsible. This is, of course, a strategic advantage to those who can successfully maintain deniability. Unfortunately, this is likely to generate a proliferation of cyber-sabotage efforts, with perhaps less concern for ethical conduct. The outcome is likely to be massive disruptions, serious harms, and social chaos—without anyone to hold accountable for it.

This leads us to our next and final point: the ethical concerns of increasing use of cyber-sabotage in military strategy.

> 3. What ethical concerns arise from the prospect of cyber-sabotage becoming a more commonly used military strategy?

In this final section, I want to briefly note some major ethical issues that are likely to arise with the increase in cyber-sabotage in the years to come.

First, recall that cyber-sabotage makes it easier to cause disruption on a mass scale. Significant and widespread damage can be done by fewer actors, from far away secure locations, without serious risk of being captured, caught, or harmed in the process. This massive scale is partly the result of an ever-increasing reliance upon technologies to support and sustain our information networks, resource systems, and other socially valuable functions.

To achieve massive disruption on a significant scale, then, almost by necessity entails that the cyber-saboteur will be causing widespread and significant harms to the civilian population. The concern here is less with the temporary disruption of, for example, social media or other recreational technologies. Rather, the concern is with systems like traffic technologies, medical systems, and, perhaps some day in the near future, self-driving vehicles. Sabotage of these systems will likely mean many civilian lives will be lost. Indeed, for the most malevolent agents, this will be their goal. Cyber-sabotage, and cyber warfare in general, therefore risk making the features that are so alarming and harrowing about methods like terrorism a more commonly used military strategy.

Moreover, given the incentives for avoiding attribution, there will be enormous risks in misattribution and mistaken retaliatory acts, which may diminish or eliminate relationships of trust between certain nations. This may, in turn, widen the scope of conflict, yielding not only more harms but also a greater number of innocents harmed.

Further, employing cyber-sabotage lowers the costs to nations—both in terms of financial costs and political capital—of achieving certain national security goals, which increases the likelihood that they will be used frequently and with perhaps less public knowledge and concern. In other words, if leaders believe that they can carry out cyber-sabotage with near political impunity, they will be less deterred from doing so.

Relatedly, the increasing availability of these methods will likely dictate military strategy. Done well, this could be a good thing: fewer "boots on the ground," fewer civilian casualties (if properly reined in), and less physical destruction. There is a risk, however, that cyber-sabotage efforts become at once more common and more expansive in nature. There are serious questions about the extent to which this will alter our global relationships, strategic partnerships, and other issues of international significance. It is difficult to predict the full extent of the ramifications; we are quickly entering unknown territory. But it is certain to raise challenging ethical issues that we cannot yet fathom. The sheer scale of uncertainty of the future of war should give us all pause.

References

Allhoff, F., A. Henschke, and B. J. Strawser, eds. *Binary bullets: the ethics of cyberwarfare*. (Oxford University Press, 2016).
Follath, E. and H. Stark. 2009. "The Story of 'Operation Orchard': How Israel Destroyed Syria's Al Kibar Nuclear Reactor." Spiegel Online International, November 2.

http://www.spiegel.de/international/world/the-story-of-operation-orchard-how-israel-destroyedsyria-s-al-kibar-nuclear-reactor-a-658663.html.

Gross, M. J. "A declaration of cyber-war." *Vanity Fair* 53, no. 4 (2011).

Lindsay, Jon R. "Stuxnet and the limits of cyber warfare." *Security Studies* 22, no. 3 (2013): 365-404.

Lucas, G. R. *Ethics and cyber warfare: the quest for responsible security in the age of digital warfare*. (Oxford University Press, 2017).

Kaplan, C. "Air power's visual legacy: Operation Orchard and aerial reconnaissance imagery as ruses de guerre." *Critical Military Studies* 1, no. 1 (2015): 61-78.

Rid, T. "Cyber war will not take place." *Journal of strategic studies* 35, no. 1 (2012): 5-32.

Sanger, D. E. *The perfect weapon: War, sabotage, and fear in the cyber age*. (Broadway Books, 2019).

Walzer, M. *Just and Unjust Wars*, fifth ed. (New York: Basic Books, 2015).

6

NOT WAR

The Ethics of "Phase Zero" Cyber Operations

Edward Barrett

Introduction

This chapter will assess ethically permissible responses to potentially or actually harmful cyber operations that do not meet the just war criteria, and thus occupy a "grey zone" between the traditional war and peace distinction. The first part will briefly examine the evolving thought in defense planning circles on the relationship between phase zero and grey zone operations. The second part will describe five relevant cases: a limitedly lethal disabling of a municipality's computer-aided dispatch (CAD) system, the placement of logic bombs within critical infrastructure, costly cyber crime and cyber espionage, the alteration of voting machine information, and COVID vaccine disinformation. The third part will explicate the ethical concepts pertinent to permissible responses—including human rights, rights forfeiture, liability to defensive and punitive harm, and *jus ad bellum* and *jus ad vim* criteria—and then apply these to each case. Given that most of the appropriate responses will not directly involve military instruments of power, the fourth part will explore the military's likely involvement in defensive and punitive responses in the grey zone.

The (Predictable) Evolution of "Phase Zero"

While Charles Krauthammer's suggestion that a "Unipolar Moment" had arrived was met with skepticism in late 1990, the stunning victory of the United States-led coalition in Operation Desert Storm had eliminated doubts by early 1991.[1] Departing from the ground-centric "AirLand Battle" doctrine that prevailed at the end of the Cold War, the operation featured coordinated airstrikes from forward-deployed bases and carriers, introduced precision-guided munitions, and routed the largest military in the Middle East in 42 days.

This triumph inspired at least three military planning initiatives that remain central to U.S deterrence and warfighting. First, the coordination that enabled Gulf War forces' rapid decision cycles would be accelerated through "network-centric" warfare prosecuted by dispersed forces—intelligence, logistics, strike and command and control—that were all linked by information systems.[2] Second and befitting a hegemon, the 1993 Bottom Up Review (BUR) deemed that the U.S. military be capable of waging two nearly-simultaneous major regional conflicts (MRCs).[3] Third and most important for our purposes, the 1993 BUR modelled future conflicts as Gulf War-like affairs that proceeded in four successive "phases." In response to a major regional power's invasion of a neighbor, the military would "halt the invasion" (Phase 1), "build up U.S. combat power in the theater while reducing the enemy's" (Phase 2), "decisively defeat the enemy" (Phase 3), and finally "provide for post-war stability" (Phase 4).[4] The current construct closely resembles the one established in 2006.[5] Phases 1 through 4 were redesignated as, respectively, "Deter," "Seize Initiative," "Dominate" and

---

1 See Charles Krauthammer, "The Unipolar Moment", *Foreign Affairs* 70:1 (1990/91). On critical responses, see Charles Krauthammer, "The Unipolar Moment Revisited," *The National Interest* 70 (2002/03).
2 On John Boyd's seminal thought on the importance of rapid decision cycles in competitive situations, see Robert Corum, *Boyd: The Fighter Pilot Who Changed the Art of War* (New York: Hachette Book Group, 2002). Early conceptualizations of network-centric warfare were William A. Owens, "The Emerging U.S. System-of-Systems," *Strategic Forum* 63 (February 1996); and Arthur K. Cebrowski and John H. Garstka, "Network-Centric Warfare - Its Origin and Future," *Proceedings* 124 (January 1998).
3 Difficulties in accomplishing this goal at acceptable budget and risk levels have been mitigated by relaxing assumptions, such as the time between MRCs and/or the time to win a first MRC. In addition to deterring and fighting MRCs, the military's mission includes "steady state operations" such as nuclear deterrence, counterdrug, peacekeeping, counterdrug and counterterrorism operations.
4 Les Aspin, *Report on the Bottom-Up Review* (October 1993), 15-16.
5 For the current version, see Joint Publication 3-0, *Joint Operations* (22 October 2018).

"Stabilize." More significantly, two phases were added. To emphasize the importance of post-conflict stability operations in the aftermath of the 2003 invasion of Iraq, a "Phase 5 - Enable Civilian Authority" was created. And on the low end of the spectrum, steady-state "Phase 0 - Shape" operations—such as theater presence, security cooperation and humanitarian engagement—were added to prevent war or ensure victory.

This emphasis on fielding conventional forces capable of quickly dominating a six-phased regional campaign has successfully deterred wars between great powers.[6] However, this success has also driven adversaries to engage in continuously harmful activities within a "grey zone" between peace and war—activities against which our forces and plans are ill-suited to defend.[7] For example, while the U.S. is technically operating in Phase 0 against China and Russia, China's constant cyber theft of intellectual property and Russia's various forms of interference in elections involve interstate rights violations that traditional Phase 0 "shaping" operations cannot address. On the other hand, the means through which the military could defensively "dominate" in these situations probably would be morally and legally impermissible, if not counterproductive.

Assuming that the military has a role to play in grey zone scenarios, a first step in assessing its relevance is to apply the Phase 0 concept only to steady-state shaping operations associated with possible MRCs, and not to grey zone operations. The venerable "military operations other than war" (MOOTW) paradigm is more apt. A second and more fundamental step is to determine which of the military's tools—kinetic and cyber—would effectively and ethically defend and deter in the grey zone. Especially because ethics, not effectiveness, is likely to be the military's limiting factor, the remainder of this chapter will explore the ethics of responding to cyber operations in the grey zone—beginning with possible cases.

---

6   China's cost-imposing strategy of fielding anti-access/area-denial (A2AD) capabilities is challenging U.S. deterrence capacities.
7   For an excellent analysis of problems with the MRC phase construct, see Paul Sharre, "American Strategy and the Six Phases of Grief," *War on the Rocks*, 6 October 2016.

Cases: Four Fictional, One Real

911 Down

It's a hot Friday evening in Chicago, and dispatchers at the city's emergency communications centers—public safety answering points (PSAPs)—are bracing to deal with the usual summer uptick in workload, augmented by COVID and protests. Budget shortfalls in most municipalities have required deep cuts to personnel and an increased reliance on computer-aided dispatch (CAD) systems, without which average response times would increase from 10 minutes to 30 minutes. Studies have estimated that for medical emergencies, this response delay would increase mortality rates by 25 percent. Additional costs would include increased suffering and health care costs, property loss and decreased productivity.

At 9 pm, Chicago's CAD system fails completely. Automatic functions that allow dispatchers to immediately map caller locations and rapidly locate and notify the closest emergency responders are lost, forcing a transition to manual modes that recently-hired dispatchers have never used. Additionally, the overwhelmed dispatchers are slow to call in information technology (IT) specialists, who eventually identify the affected server, determine that it is unsalvageable, and order a replacement that will not be installed for at least a week. Anticipating a full week of manual operations, the mayor approves overtime pay and requires all dispatchers to work double shifts. Thankfully, the CAD system is repaired by Thursday. But tragically, the one-week loss of capability is estimated to have resulted in eight additional deaths.

The next Friday, New York City's CAD system goes down, although national news generated by the Chicago failure results in a speedy IT specialist response and salvageable server. Nevertheless, the city's dispatchers are forced to operate manually until Monday morning, resulting in an estimated five additional deaths. The Friday after, Los Angeles' system fails. Bracing for a long weekend, one of its first responders remarks to a colleague, "A coincidence? Right." Laying down the last newspaper she'll read for days, she glances at a headline: "China and Russia Protest as U.S. Doubles Arms Sales to Taiwan and Ukraine."

Questions:

- Do such harms justify a lethal response?

- What degree of certainty about an attacker's identity is required to respond?

Bombs in the Grid

Since the shockingly sophisticated CrashOverride cyber attack on Ukraine in 2016, the North American Electric Reliability Corporation (NERC)—the industry group responsible for power grid security in the U.S. and Canada—has been working overtime. CrashOverride was a malicious piece of software (malware) hiding in a computer that was part of the Supervisory Control and Data Acquisition (SCADA) network allowing operators to monitor and control Kiev's electrical infrastructure. At a preprogrammed time, the logic bomb automatically activated and popped circuit breakers at substations, creating winter darkness in western Kiev. Although technicians were able to detach the breakers from the computer and restore power in just over an hour, the malware's capabilities, reusability and lack of fingerprints raised alarms among infrastructure security experts.

Five years after the attack, NERC's work with industry and governmental experts has produced CrashOverride detection software, although the ability to remove the malware is still under development. As the detection software is finally installed on the East Coast during an historically cold winter, a chilling discovery is made: the malware is found in the electrical grid SCADA systems of nearly every urban area from Boston to Washington. Experts at the National Security Agency (NSA) and Cyber Command (CYBERCOM) quickly determine that in addition to tripping circuit breakers, it can destroy electrical generators and thus cause a mass casualty event. Technicians immediately disable SCADA networks. A week later, cyber security specialists learn that the malware's launcher component contained no activation time and date, and trace the malware installations to hackers thought to be affiliated with the Russian military's "GRU" intelligence agency.

Questions:

- Is the installation of such logic bombs an act of war?
- What if the malware has been programmed to activate automatically at a specific time?
- Assuming that a link with the Russian government cannot be verified, does the U.S. have the right to retaliate against the hackers located within Russia's borders?

A Thousand Cuts

The economic costs of cyber crime and cyber espionage by foreign actors include: the loss of commercial intellectual property; direct financial loss; lost negotiating strategies; cyber security costs; opportunity costs, such as additional spending to offset losses of military technology and jobs; and reduced real wages and employment. For years, these costs have been tolerable, resembling those of car crashes and pilferage—approximately 1 percent of GDP (Gross Domestic Product) and one-third of a percent in employment. Despite these drags on the economy, GDP growth has been averaging 3 percent, unemployment has been at historic lows, and real wages for all socioeconomic groups have been growing.[8]

However, the evidence and costs of malicious cyber activity have skyrocketed over the past three years. A Chinese iPhone clone—"cPhone"—has reduced Apple's share of the global smartphone market from 15 percent to 7 percent, requiring layoffs of 20,000 employees. Cyber crime and cyber espionage are now estimated to annually shave 5 percent off the GDP, accounting for a GDP that is contracting by 1 percent annually. Unemployment has jumped from 4 percent to 12 percent, wages have fallen, and an already-underfunded entitlement system is facing new pressures. On the other hand, China's GDP is growing 10 percent annually, and its defense spending has increased from 2 percent to 4 percent. When the U.S. and Chinese defense budgets are compared using purchasing power parity (PPP), and given that only two-thirds of U.S. forces are relevant to deterrence and warfighting in the Pacific, regional stability is questionable. There are rumors that Japan is consulting with Israel about nuclear deterrence forces.

At a special meeting of the Council of Economic Advisors and senior U.S. intelligence and defense leaders, the President wants the facts. Intelligence experts insist that losses from foreign agents based in the U.S. are relatively low, that our cyber attribution capabilities are reliable, and that the majority of the harm originates from "Unit 61398" of the Chinese military's cyber forces and that unit's non-state clients are in mainland China. The Chinese deny wrongdoing, and our passive defenses and deterrence by denial are increasingly ineffective. The President turns to the Secretary of Defense and Director of National Intelligence and says, "It's time to take the fight to China. What are our options?"

---

8 For an overview of these costs, see Center for Strategic and International Studies, *The Economic Impact of Cybercrime and Cyber Espionage* (2013).

Questions:

- Do these harms justify a lethal response?
- If not, which sublethal responses would be justified?
- Who would be the legitimate targets of any response?

Election Surprise

Alarmed by destabilizing controversies afflicting the past two presidential elections, federal and state governments have cooperated to mandate and implement more standardized and effective vote integrity measures. Nationwide, voter rolls have been verified, voter access and identification procedures improved, ballot control and counting procedures tightened, and electronic voting machines better certified. Thankfully, the country has a new holiday to encourage in-person voting: National Voter Day. To mitigate foreign influence operations, federal and state governments are scrutinizing campaign contributions and—in conjunction with all forms of media—information sources. Largely due to these efforts, voter turnout set new records and a close presidential election was not disputed.

Two years into the new administration, the Director of National Intelligence (DNI) asks the President to convene a classified meeting of the National Security Council. The DNI delivers the harrowing news: the President did not win the last election. Ironically, the firm that provided cyber security software to every electronic voting machine was infiltrated by Russian hackers. Malware was dispersed through a software update just prior to the election, and enough votes were shifted from the vehemently anti-Russia candidate to change the winner of both the popular vote and electoral college.

The nation is at peace, a favorable trade agreement with China has just been signed, unemployment and crime are low, the GDP is growing at 5 percent annually, real wages are broadly rising, and the administration has just shepherded through Congress landmark legislation on immigration, healthcare and entitlement reform. The President's approval rating is 65 percent—as high as Eisenhower's in 1955. The President turns to the DNI: "Are you sure?" His response: "Absolutely."

Questions:

- Does this level of election interference constitute an act of war?
- What if the interference were discovered 20 years later?

Vaccine Lies

Our final and real case involves COVID-19 vaccines. In many ways, the U.S. response to the virus was relatively lackluster. Compared to Taiwan and South Korea, testing, tracing and quarantine efforts were poor; and compliance with distance and mask protocols was mixed. But the U.S. excelled in one respect: vaccine discovery, production and distribution. Its companies—Moderna, Pfizer and Johnson & Johnson—produced the first, safest and most effective vaccines; and by mid-May 2020, the U.S. was near the top in doses administered per 100 people.[9]

Predictably, hackers and intelligence agencies from Russia and China attempted to steal vaccine research, production and distribution information, but were mostly unsuccessful.[10] However, these organizations were more successful in spreading disinformation about the vaccines' costs, safety and efficacy. Leveraging preexisting vaccine skepticism and using a variety of media, these states amplified existing critiques and created new ones in order to increase the market share of Russian and Chinese vaccines (Sputnik X and Sinovac), exacerbate U.S. social divisions, and reduce the U.S. vaccination rate.[11]

The effects of vaccine rate reductions on COVID cases and deaths could be serious, especially if they prevented herd immunity. A study by the Imperial College London estimated death rates in highly skeptical countries such as France would be 8.7 times higher than the rate in places with ideal vaccination rates of 98 percent, while rates in moderately skeptical Germany and low hestitancy U.K. would be 4.5 and 1.3 times higher, respectively. Overall, modest levels of hesitancy could add 236 deaths per million people over a two year period.[12]

---

9 CNN (2021), "Tracking COVID Vaccinations Worldwide", available at https://edition.cnn.com/interactive/2021/health/global-covid-vaccinations/.
10 BBC (2020), "Coronavirus: Hackers Targeted Covid Vaccine Supply "Cold Chain"", available at https://www.bbc.co.uk/news/technology-55165552.
11 *Wall Street Journal* (2021), "Russian Disinformation Campaign Aims to Undermine Confidence in Pfizer, Other Covid-19 Vaccines, U.S. Officials Say", available at https://www.wsj.com/articles/russian-disinformation-campaign-aims-to-undermine-confidence-in-pfizer-other-covid-19-vaccines-u-s-officials-say-11615129200; *Time* (2021), "Meet the Russian 'Information Warrior' Seeking To Discredit COVID-19 Vaccines", available at https://time.com/5948017/news-front-covid-19-information-war/; Reuters (2021), "Russia, China Sow Disinformation to Undermine Trust in Western Vaccines: EU, available at https://www.reuters.com/world/china/russia-china-sow-disinformation-undermine-trust-western-vaccines-eu-report-says-2021-04-28/.
12 Bloomberg (2021), "Vaccine Skepticism Risks Increasing Covid Mortality Ninefold", available at https://www.bloomberg.com/news/articles/2021-03-25/vaccine-skepticism-could-increase-covid-mortality-up-to-ninefold.

Questions:

- Could vaccine disinformation warrant a lethal response?
- Why would a sublethal response to vaccine disinformation usually be more appropriate?

Ethical Analyses

As other pieces in this volume have explored, the venerable just war tradition has been the *locus classicus* of ethical thought on permissible responses to unjustified harm between states. Gradually, because of a 500-year human rights revolution in the West, our understanding of the just war criteria and retributive justice have become more explicitly anchored to the logic of human rights. Most of the *jus ad bellum* and *jus in bello* criteria can be grounded in the assertion that persons normally possess an inviolable right to life.[13] In the context of war, the existence of human rights means not only that aggression is impermissible, but also that defensive uses of lethal force are impermissible unless an aggressor has forfeited their right to life. This principle of forfeiture has been developed by revisionist philosophers under the rubric of "liability to lethal defensive harm."

While the preconditions for being liable to intentional harm are debated, I would argue that they include the aggressor's threat level and culpability, and a response's effectiveness and necessity. Grave and culpable rights violations such as attempted murder compromise one's right to life to the requisite degree and are the basis for the *ad bellum* just cause criterion. Effectiveness and necessity are aspects of liability because even murderous acts do not eliminate a person's potentialities and worth. The effectiveness precondition generates the *ad bellum* reasonable chance of success criterion, while the necessity requirement creates the *ad bellum* last resort and the *in bello* prohibition against killing combatants unnecessarily.[14] The existence of human rights in war also gives rise to the discrimination and proportionality criteria. Because civilian bystanders in a war zone have forfeited none of their rights and are not liable to any harm, defenders must target only liable combatants. And when doing so, they must treat

---

13 For the purposes of this chapter, the *ad bellum* criteria are right intention, proper authority, just cause, last resort, reasonable chance of success, and broad proportionality; and the *in bello* criteria are discrimination, necessity and broad proportionality.
14 I would argue that the reasonable chance and last resort criteria should also be applied *in bello*.

innocents with due care. In accordance with the principles of double intention and effect, soldiers must incur acceptable risks to avoid killing bystanders, and then cause only a proportionate amount of unintentional but foreseeable harm to them.[15] Let's return to our cases to explore the applicability of these moral requirements.

911 Down

This scenario highlights two ethical challenges associated with cyber warfare ethics. The first—which is germane to kinetic and cyber attacks—concerns whether a just cause for war exists. Given the unjustified loss of life caused by the CAD system destruction and damage, a lethal response would be narrowly proportionate and thus permissible.[16] However, given the relatively low number of deaths, any defensively-oriented lethal response would have to be limited. Attempting to address such situations, scholars have developed a *"jus ad vim"* ("justice before force") framework to morally assess violent "measures short of war."[17] While they argue that the traditional criteria fail to account for relevant ethical requirements, Helen Frowe has argued—correctly, I believe—that the traditional criteria, properly understood, incorporate all of the concerns allegedly addressed only by an *ad vim* framework.[18] Therefore, a rightly-motivated state could respond lethally, but only if all the *ad bellum* and *in bello* criteria were met. And even if a lethal strike would be effective and necessary, the wide proportionality requirement to achieve a net benefit would certainly limit the response.

---

15 On the principle of double intention, see Michael Walzer, *Just and Unjust Wars: A Moral Argument with Historical Illustrations* (Basic Books, 1977), chapter 9.
16 A response is narrowly proportionate when the harm done to the wrongdoer is commensurate with the importance of the rights being defended. A response is widely proportionate when the harm done to those who are not liable to be harmed is not excessive in relation to the response's good effects. See Jeff McMahan, "Liability, Proportionality, and the Number of Aggressors," in Seth Barzagan and Samuel Rickless, *The Ethics of War: Essays* (Oxford, 2017).
17 See Michael Walzer, *Just and Unjust Wars: A Moral Argument with Historical Illustrations* (Basic Books, 2006), pp. xv-xxvi. Developments of *jus ad vim* include: See Daniel Brunstetter and Megan Braun, "From *Jus ad Bellum* to *Jus ad Vim*: Recalibrating Our Understanding of the Moral Use of Force," *Ethics & International Affairs* 27:1 (2013); Daniel Brunstetter and Megan Braun, "Rethinking the Criterion for Assessing CIA-Targeted Killings: Drones, Proportionality and *Jus ad Vim*," *Journal of Military Ethics* 12:4 (2013); and Brandt S. Ford, "*Jus ad vim* and the Just Use of Lethal Force-Short-of-War," in Fritz Allhoff, Nicholas Evans, and Adam Henschke, eds., *Routledge Handbook of Ethics and War* (Routledge, 2013).
18 Helen Frowe, "On the Redundancy of *Jus ad Vim*: A Response to Daniel Brunstetter and Megan Braun, *Ethics & International Affairs* 30:1 (2016), 120-123.

Assuming that a defensive use of force is permissible, perpetrator attribution adds a second challenge to cyber cases. Technical aspects of computers and the internet can undermine certainty about a wrongdoer's location, equipment, personal identity, and/or institutional affiliation—and can even implicate innocent parties. In the CAD system case, circumstantial evidence—motive, means, and opportunity—points to many parties, not just China and Russia. Especially when responding lethally (as uneasiness with the domestic capital punishment attests), probability is not enough.[19] Certainty is required. Since circumstantial evidence does not adequately establish certainty and may be all that is available in cyber attack cases, passive defenses against, and denial-based deterrence of, future incidents may the only legitimate responses.[20]

Bombs in the Grid

At first glance, planting inert logic bombs across the East Coast electrical grid seems less troubling than the *911 Down* and Ukraine 2016 cases, given that the harms in the latter two situations were actual. But the potential harms in the East Coast case are massive and even existential. Especially if there were widespread destruction of generators in winter, thousands could perish from cold, starvation and violence. And the loss of nuclear warning and command and control systems would render the U.S. vulnerable to a nuclear attack.

However, despite these extraordinary stakes, nothing harmful has happened yet, and any harmful response would thus have to satisfy the demanding requirements of anticipatory defense. One could argue that wrongdoers forfeit rights and incur liability when actively preparing to harm, which is why conspiracy is a crime. But because active preparation is difficult to verify internationally, imminent aggression—for example, mobilized forces on one's border—traditionally has been considered the valid indicator of liability. Unfortunately, and unlike nuclear attacks, there is no way to know when logic bomb attacks are imminent (unless they are preprogrammed and one is privy to the activation time). Therefore, when imminence is unknowable, certainty about active preparation to cause

---

[19] For a utilitarian defense of responding when only 90 percent certain, see Randall Dipert, "The Ethics of Cyberwarfare," *Journal of Military Ethics* (2010) 384-410. Although I disagree with this suggestion, Dipert's article was the first on cyber warfare ethics and remains definitive.

[20] States publicly inflate their attribution capabilities in order to maximize deterrence.

grave harm would justify anticipatory self-defensive measures—if the other *ad bellum* criteria, including necessity/last resort, were met.[21] But a warning is in order: this form of anticipatory defense is morally hazardous for epistemic and bureaucratic reasons. Obtaining certainty about active preparation is notoriously difficult, as is creating organizations where intelligence analysts' assessments are correctly interpreted and available. These two problems highlight the importance of passive defenses in an era of difficult-to-deter and stealthy threats.

Assuming that all of these ethical hurdles are cleared, and that Russian government collusion with the hackers cannot be verified, how could the U.S. justify a strike against non-state actors within Russian borders? Stated differently: at what point does an attacked state's right to defend itself trump the sovereignty of a state that is the geographic source, but not the cause, of a lethal attack? Although sovereignty is a legal concept, its ethical foundation is the right of individuals to associate for political purposes and the duty of others to not interfere. However, sovereignty is also accompanied by responsibilities, internal and external. Just as states that harm their own citizens can lose their right to non-interference, creating a permission or even obligation to intervene, states who are unwilling to protect other states' citizens from threats emanating from or traversing their jurisdictions, or who are unable to do so and refuse necessary assistance, also lose their right of non-intervention.

Finally, what if a fully autonomous system could detect and thwart an imminent attack by state or non-state actors on critical infrastructure, eliminating the need for morally hazardous forms of anticipatory defense? While automated passive defenses would be ethically unproblematic, automated cyber countermeasures that strike network targets inside another state should not be used prior to an existing conflict. While operationally advantageous, such systems should not be allowed to subvert the ethical requirement that legitimate political authorities decide whether to initiate conflict.

A Thousand Cuts?

Can an accumulation of harmful events, which individually would not be a sufficient reason to use lethal defensive force, constitute a just cause?

---

21 In many cases, simply removing the infected computers from the grid would eliminate the threat.

Although much cyber threat discussion has focused on the possibility of catastrophic damage to critical infrastructure, the more pressing issue is whether the constant economic costs that undermine national prosperity could be considered a "death by a thousand cuts."[22]

Ultimately, situational details will determine which responses are ethically appropriate. As previously argued, persons forfeit their right to life and therefore become liable to lethal force only when culpably employing, or planning to employ, means known to be capable of murdering or destroying/stealing life-sustaining property. So on the one hand, culpably attempting or planning to cause low-level harm as part of an existentially-threatening campaign would create a liability to lethal kinetic or cyber force. However, in these cases, uses of force would be preventive, and thus subject to the moral concerns about anticipatory self-defense outlined earlier.

On the other hand, constant cyber intrusions that combine to merely reduce living standards do not justify lethal responses, which would be disproportionate. By analogy, killing a professional pickpocket—even if doing so were effective and necessary for defense against theft—would be disproportionate and thus unjustified. But serial thievery is nevertheless unjust, and in an international system lacking effective due process and penal mechanisms, victim states have the right to respond with sublethal defensive measures. Unfortunately, the effectiveness of such measures might be limited. Incarceration might be impossible, and passive defenses and sublethal countermeasures might not fully degrade harmful capabilities. However, the requirement to respond sublethally creates two possibilities with defensive ramifications that do not obtain for lethal defensive harms: inflicting proportionate punitive harm, and targeting indirect participants. Let's briefly discuss each of these.

The traditional justifications for punitive harm—such as capital punishment of safely imprisoned criminals, austere incarceration conditions, and non-compensatory fines—are retributive and consequentialist. Retributivists argue that harming wrongdoers is justified because they deserve it; their suffering is good in itself. Consequentialists justify punitive harm as an effective and necessary means to attaining rights-related benefits, including the specific deterrence of a perpetrator from engaging in future wrongdoing, the reform of the perpetrator, and the

---

22 The level of this theft-based harm is disputed, in part because companies are reticent to report it.

general deterrence of other would-be wrongdoers. Although the relative merits of these justifications are still debated, I offer three assumptions that are restrictive and resonate in liberal societies. First, human rights and their forfeiture require that intentionally inflicted suffering serve a purpose, thus ruling out desert-based retributive punishments.[23] Second, human rights and their forfeiture require that even wrongdoers not be used as an example-setting means of preventing possible wrongdoing by unspecified others, which rules out the general deterrence justification.[24] Third, the remaining and legitimate consequentialist purposes of punishment are specific deterrence and reform.

Because lethal harm cannot accomplish these two legitimate purposes, capital punishment *qua* punishment and punitive war are impermissible. But sublethal harm can do so, and therefore may be used both defensively and punitively if proportionate, effective and necessary. Given that incarceration and subsequent reform are unlikely when dealing with foreign aggressors, the primary purpose of such punitive harm would be specific deterrence. Accordingly, punitive measures would seek not to degrade harmful capabilities, but instead to alleviate malevolent intentions by degrading the wrongdoer's standard of living—especially personal property and reputation. And ideally, such measures would have a defensive effect and render further harms unnecessary.[25]

The second possibility created by the requirement to respond sublethally is the targeting of indirect participants. The ethics and law of war rightly draw a bright line around civilians, who are at worst only

---

[23] Retributivism has been defended as a means of emphasizing the validity of a society's morality—a way of "planting the flag of moral truth." However, this is a consequentialist general deterrence argument, and therefore problematic from my perspective. Alternatively, retributivism has been defended as a means of securing another consequence: a victim's right to "expressively defeat" the indignity. See David Luban, "War as Punishment," *Philosophy and Public Affairs* 39:4 (2012). Luban's article is a seminal and sympathetic treatment of the historical shift away from punitive war, however. While he defends retributivism, he argues that because of human psychology and the lack of an impartial third party, punitive war in practice is revengeful and unjust because it over-harms (citing Suarez).

[24] Although one could argue that a right to not be intentionally punished in order to deter others could be overridden by benefits of general deterrence, I suspect that these benefits would be either difficult to prove or insufficient to override the right. However, general deterrence might be a side effect of permissible defensive and punitive harm.

For an important duty (to the victim)-based justification of punishing the guilty for the purpose of general deterrence, see Victor Tadros, *The Ends of Harm: The Moral Foundations of Criminal Law* (Oxford: Oxford University Press, 2011). For an effective critique of Tadros' arguments, see Kimberly Kessler Ferzan, "Rethinking *The Ends of Harm*," *Law and Philosophy* 32 (2012).

[25] Specific deterrence is ultimately defense through different means.

indirectly responsible for unjustified lethal harms and thus not liable to be killed. But indirect participants in unjustified lethal or sublethal attacks may be liable to sublethal harms—a difference that comports with criminal law. Therefore, non-cooperative political leaders of the territory where a wrongdoer resides, as well as civilian accomplices such as financiers, would surely be liable to some form of defensive and punitive sublethal harms. The combination of these countermeasures and punishments might be enough to reduce the harm, and even coerce states to cooperate and accomplices to desist.

Election Surprise

Determining a proportionate response to this act—a cyber attack that changes the result of a presidential election—requires clarity about what rights have been violated. Because violence was neither a cause nor effect of the "regime change," one could argue that the perpetrators would not be liable to a lethal response. Given how liberally and competently the current administration has performed, the only possible harm seems to be in publicly disclosing the interference, which could result in domestic and/or international strife. However, any strife would be rooted in the widespread violation of important voting rights. One can argue that like other rights, the right to participate in the selection of one's political leaders is a precondition to human flourishing. A well-designed democracy is regarded as the most reliable means of protecting basic human rights, and political participation helps cultivate important virtues. And although the cyber attack disenfranchised only a relatively small percentage of voters and resulted in an ethically legitimate regime, it also could undermine confidence and participation in politics, and eventually lead to an erosion of other social and political rights. This danger accounts for the severity of punishments for even small-scale voter fraud, which is a felony.

But for ethically sound reasons, voter fraud is not a capital crime. As noted earlier, responses must be proportionate; the severity of the rights violation determines the harm permitted in response. It is possible that voter fraud could intentionally install a tyrannical regime, in which case the fraudsters and tyrants might be liable to lethal defensive force. But in most cases, including the one at hand, the stealing of votes is not life threatening, and therefore justifies only a sublethal response. However, as argued above, sublethal responses to serious election interference may be defensive and punitive, and target direct and indirect participants.

Finally, what if the election interference were initially undetected, unattributed or covered up, and its perpetrators finally known 20 years later? Assuming that no additional attacks of the same sort have occurred, any sublethal response at that late point could not be justified as a defensive act. Additionally, as it would be difficult to argue that the perpetrators need to be deterred anymore, punishment for the purpose of specific deterrence seems unjustified. But even if defensive and punitive harm are prohibited, the perpetrators owe compensation for damages and may be coercively punished for refusing to compensate, and state actors who protect them may be punished for doing so.

Vaccine Lies

Given the additional deaths resulting from vaccine skepticism, there seems to be just cause for a lethal response. However, assuming reliable attribution, at least three facets of the situation militate against responding lethally. First, because there are so many possible causes of one's opinions, establishing a causal link between the disinformation and skepticism would be difficult. Second, even if a link can be established, the degree of causation is likely to be low, in turn reducing the harm to which the perpetrator would be liable. Especially in cases where wrongdoers are merely amplifying existing critiques among already skeptical individuals, the culpability for additional deaths would be extremely low.

Third, even if we assume that disinformation was solely created by a perpetrator and heavily influenced a recipient who was not already skeptical, the recipient would bear at least some of the responsibility for his erroneous viewpoint. Unlike coercion, manipulation does not override the will. And adults in a free society have opportunities to inform themselves. Unless manipulators have somehow eliminated the possibility of accessing alternative viewpoints, those who are manipulated are to some degree culpable for their ignorance, again reducing the wrongdoer's liability.

The Military in Grey Zone Operations

Finally and briefly, I wanted to end where we began—with operational considerations. With this chapter's discussion of the ethics of grey zone responses in the background, what should be the military's defensive and punitive roles in the grey zone?

On the one hand, when limited lethal defensive responses are appropriate, they normally should be handled by the military. Because small-scale kinetic responses create more possibilities for low-visibility clandestine and covert operations, they raise a couple of concerns that require attention. First, all intelligence community (IC) assets that conduct strike operations must be trained and held to the military's high standards, an imperative that might require cultural changes in these organizations. Second, whether conducted by the IC or military, limited lethal operations will require adequate legislative oversight to ensure compliance with the *ad bellum* and *in bello* criteria.

On the other hand, as our cases have illustrated, because most defensive and punitive grey zone responses against direct and indirect participants should be sublethal, non-military instruments of power—diplomatic, informational and economic—will be the most appropriate tools. Some of these tools will directly involve the military. For example, the U.S. Cyber Command might execute information and economic operations that occur in cyber space. But because other U.S. departments and agencies—especially State, Treasury, Commerce, Homeland Security and the intelligence agencies—will take the lead in these operations, the military will often have a supporting role in a less martial and increasingly complicated "interagency process."

References

Aspin, L. *Report on the Bottom-Up Review* (October 1993).
BBC (2020), "Coronavirus: Hackers Targeted Covid Vaccine Supply "Cold Chain"', https://www.bbc.co.uk/news/technology-55165552.
Bloomberg (2021), "Vaccine Skepticism Risks Increasing Covid Mortality Ninefold", https://www.bloomberg.com/news/articles/2021-03-25/vaccine-skepticism-could-increase-covid-mortality-up-to-ninefold.
Brunstetter D. and M. Braun, "From *Jus ad Bellum* to *Jus ad Vim*: Recalibrating Our Understanding of the Moral Use of Force," *Ethics & International Affairs* 27:1 (2013).
Brunstetter D. and M. Braun, "Rethinking the Criterion for Assessing CIA-Targeted Killings: Drones, Proportionality and *Jus ad Vim*," *Journal of Military Ethics* 12:4 (2013).
Cebrowski, A.K. and J. H. Garstka, "Network-Centric Warfare - Its Origin and Future," *Proceedings* 124 (January 1998).
CNN (2021), "Tracking COVID Vaccinations Worldwide", https://edition.cnn.com/interactive/2021/health/global-covid-vaccinations/.

Corum, R. *Boyd: The Fighter Pilot Who Changed the Art of War* (New York: Hachette Book Group, 2002).
Dipert, R. "The Ethics of Cyberwarfare," *Journal of Military Ethics* (2010).
Ford, B.S. "*Jus ad vim* and the Just Use of Lethal Force-Short-of-War," in Fritz Allhoff, Nicholas Evans, and Adam Henschke, eds., *Routledge Handbook of Ethics and War* (Routledge, 2013).
Frowe, H. "On the Redundancy of *Jus ad Vim*: A Response to Daniel Brunstetter and Megan Braun, *Ethics & International Affairs* 30:1 (2016).
Kessler Ferzan, K. "Rethinking *The Ends of Harm*," *Law and Philosophy* 32 (2012).
Krauthammer, S. "The Unipolar Moment", *Foreign Affairs* 70:1 (1990/91).
Krauthammer, S. "The Unipolar Moment Revisited," The National Interest 70 (2002/03).
Luban, D. "War as Punishment," *Philosophy and Public Affairs* 39:4 (2012).
McMahan, J. "Liability, Proportionality, and the Number of Aggressors," in Seth Barzagan and Samuel Rickless, *The Ethics of War: Essays* (Oxford, 2017).
Owens, W.A. "The Emerging U.S. System-of-Systems," *Strategic Forum* 63 (February 1996).
Reuters (2021), "Russia, China Sow Disinformation to Undermine Trust in Western Vaccines: EU, https://www.reuters.com/world/china/russia-china-sow-disinformation-undermine-trust-western-vaccines-eu-report-says-2021-04-28/.
Sharre, P. "American Strategy and the Six Phases of Grief," *War on the Rocks*, 6 October (2016).
Tadros, V. *The Ends of Harm: The Moral Foundations of Criminal Law* (Oxford: Oxford University Press, 2011).
Time (2021), "Meet the Russian 'Information Warrior' Seeking To Discredit COVID-19 Vaccines", https://time.com/5948017/news-front-covid-19-information-war/.
Wall Street Journal (2021), "Russian Disinformation Campaign Aims to Undermine Confidence in Pfizer, Other Covid-19 Vaccines, U.S. Officials Say", https://www.wsj.com/articles/russian-disinformation-campaign-aims-to-undermine-confidence-in-pfizer-other-covid-19-vaccines-u-s-officials-say-11615129200.
Walzer, M. *Just and Unjust Wars: A Moral Argument with Historical Illustrations* (Basic Books, 1977).
Walzer, M. *Just and Unjust Wars: A Moral Argument with Historical Illustrations* (Basic Books, 2006).

7

ETHICS OF MILITARY CYBER SURVEILLANCE

Peter Lee

Introduction

Cyber capabilities offer governments and militaries previously unimaginable opportunities across the traditional domains of land, sea and air, as well as providing a distinct domain of operations. Technologically advanced states, especially major powers like the U.S., China and Russia, are investing huge human and financial resources to establish or maintain cyber dominance, with the associated political, military, social and economic advantages and implications it brings.[1] Rid argues that "[c]yber war will not take place," at least not as a discreet activity separated from any other kind of war, on the basis that no "single cyber offense on record constitutes an act of war on its own."[2] However, he also recognises that cyber offers unique opportunities within long-established activities associated with war: "sabotage, espionage, and subversion."[3] The focus of this chapter is linked to all three of these activities, though perhaps most commonly seen as part of espionage: military cyber surveillance and the ethical challenges involved.

In a military context, cyber provides the ability to surveil enemies, potential enemies and even allies—including their Defence organisations and private military sub-contractors—at relatively low cost and similarly low physical risk. Techniques include the use of malicious software, spearphishing (tricking targets into sharing passwords, usernames or

---

1 Lyu Jinghua, "What Are China's Cyber Capabilities and Intentions?", Carnegie Endowment for International Peace, 1 April 2019, https://carnegieendowment.org/2019/04/01/what-are-china-s-cyber-capabilities-and-intentions-pub-78734.
2 Thomas Rid, "Cyber War Will Not Take Place," *Journal of Strategic Studies* 35, no. 1 (1 February 2012): 5–32, https://doi.org/10.1080/01402390.2011.608939.
3 Rid, 6.

other sensitive information)—and other hacking approaches—"to hack into sensitive financial systems, conduct massive data breaches, spread ransomware, attack critical infrastructure, and steal intellectual property."[4] The U.S. has accused China of "cyber-enabled economic espionage and trillions of dollars of intellectual property theft" which has indirect defence implications for both sides.[5]

The gathering of intelligence through monitoring of key infrastructure, enemy activities, or leadership plans and intentions, can provide both strategic advantage and the operational ability to anticipate major decisions and actions. These can take place in peacetime, periods of escalating political and military tension, or during armed conflict and war. They can also occur directly between one state and another, indirectly through the employment of non-state groups, or cyber intrusions can be routed through third party states for the purpose of avoiding detection or providing a veneer of deniability on the part of the aggressor. The surveillance considered in this chapter will be limited to that aspect of espionage which involves penetrating "an adversarial system for purposes of extracting sensitive or protected information," in this case the cyber domain.[6] Cyber surveillance—or apparent cyber attacks whose purpose is surveillance or data gathering—can be military, military-related, civilian, or a combination of the three. It can also take place outside times of war or armed conflict. As Beard observes, "Intelligence is not, and ought not to be, restricted to times of war."[7] An example which spans these categories include the China-linked CactusPete Advanced Persistent Threat (APT) which targeted military and financial organisations in Eastern Europe.[8] Kaspersky used a Threat Attribution Engine to analyse the code and assess it against known threats.[9] Their analysis discovered "300 samples appeared between March 2019 and April 2020—about 20 samples per month."[10]

---

4 Donald Trump, "National Cyber Strategy of the United States of America" (Washington D.C.: The White House, September 2018), 11, https://www.whitehouse.gov/wp-content/uploads/2018/09/National-Cyber-Strategy.pdf.
5 Trump, 2.
6 Rid, "Cyber War Will Not Take Place," 20.
7 Matthew Beard, "Just War, Cyberwar and Cyber-Espionage," in *Ethics and the Future of Spying* (Oxford: Routledge, 2016), 108.
8 Pierluigi Paganini, "Chinese APT CactusPete Targets Military and Financial Orgs in Eastern Europe," Security Affairs, 14 August 2020, https://securityaffairs.co/wordpress/107128/apt/catuspete-updated-backdoor.html.
9 Kaspersky, "Kaspersky Researchers Uncover New Targeted Campaign against Financial and Military Organizations."
10 Kaspersky.

Barrett has used a just war framework to address some ethical aspects of cyber warfare and provides a helpful reminder that "states have both the right and the duty to defend their citizens against life-threatening attacks that will occur without warning at an unknowable time."[11] The just war *jus ad bellum* and *jus in bello* criteria he specifies, respectively, include "legitimate authority, just cause, right intention, last resort, reasonable chance of success, and proportionality…[and] discrimination, military necessity, and civilian due care."[12] However, the nature of surveillance and characteristics of the cyber domain—such as the CactusPete example above—do not lend themselves to a comprehensive ethical analysis using only a just war framework. This chapter therefore supplements just war analysis, where it can be applied, with wider deontological and consequentialist ethics considerations. The former is concerned with process—rules, obligation to comply with moral laws, and means[13]—while the latter is concerned with outcomes or consequences.[14]

This chapter therefore examines aspects of military cyber surveillance ethics in both peacetime and wartime, and across military and military-related actors. Discussion of these elements is broken down into three sections and proceeds by considering the broader context within which related ethical judgements about cyber surveillance are made, including sovereignty and international law. Section two examines the ethical implications of pre-emptive versus preventive military cyber surveillance, followed by a discussion of the difficulty and importance of attribution of cyber surveillance. The final section introduces the ethics concept of *jus ad vim*—force-short-of-war—and uses recent cyber intrusions to explore the challenge of ethical analysis and appropriate political response.

Ethics, Sovereignty and Law

The practical operation of cyber surveillance raises new ethical questions on old themes. To what extent does it violate state sovereignty? Is it the equivalent of watching across a border with binoculars—or flying a

---

11 Edward T. Barrett, "Warfare in a New Domain: The Ethics of Military Cyber-Operations," *Journal of Military Ethics* 12, no. 1 (April 2013): 7, https://doi.org/10.1080/15027570.2013.782633.
12 Barrett, 14.
13 Richard Norman, *The Moral Philosophers: An Introduction to Ethics* (Oxford: Oxford University Press, 1998), 70ff.
14 Norman, 160ff.

remotely piloted aircraft on the edge of another state's airspace and training a powerful camera into their territory? Or is it like putting an intelligence agent on the ground in someone else's sovereign territory? Ethical judgement is also shaped by—but not the same as—the extent such activities violate, or otherwise, international law. All of these questions are made more complex by disputes over the nature and extent of state sovereignty.

Bartelson describes how the tenuous, contested concept of sovereignty provides the basis of both statehood and the international system itself, "hovering somewhere between them, but residing in neither."[15] Elshtain attaches conditions to sovereignty and statehood: "[W]e cannot assume that a nation-state is sovereign until it demonstrates its ability to be independent from the protection of another state, to treat its citizens decently, and to foster a vibrant civil society: sovereignty as responsibility."[16] A central element of sovereignty and statehood is the requirement for internal security to protect the common life of citizens *within* a state while at the same time providing external security and protecting that common life from attacks from other states or non-state groups.

This view holds implications for physical, as opposed to cyber, intervention and illustrates one approach to statehood and sovereignty in what is known as the "'unwilling or unable' test."[17] Consider the following example. The U.S. has used Reaper and Predator drone strikes against terrorists and terrorist groups in a number of states on the basis that it has been a victim of terrorist attack and its own sovereignty violated. The U.S. claims a right "to engage in extraterritorial self-defense where the host is either unwilling or unable to take measures to mitigate the threat posed by domestic non-state actors, thereby circumventing the need to obtain consent from the host state."[18] Where the host state has given approval for drone strikes in its territory there is no violation of sovereignty. Further, if the state is "unwilling or unable to suppress the threat posed by the individual being targeted"[19]—part of what Elshtain calls the rights and responsibilities of sovereignty—then there is no sovereignty to violate in the first place.

In the post-9/11 years, sovereignty and state rights have been increasingly prominent in international discourse around war, intervention

---

15 Jens Bartelson, *A Genealogy of Sovereignty* (Cambridge: Cambridge University Press, 1995), 46.
16 Jean Bethke Elshtain, *Sovereignty* (New York: Basic Books, 2008), 228.
17 Gareth D. Williams, "Piercing the Shield of Sovereignty: An Assessment of the Legal Status of the 'Unwilling or Unable' Test," *University of New South Wales Law Journal* 36, no. 2 (2013): 625.
18 Williams, 625.
19 Rosa Brooks, "Drones and the International Rule of Law," *Journal of Ethics and International Affairs* 28 (2014): 97.

and the pursuit of national self-interests. Cyber has been part of each of these activities, starting with its use by Russia as one of several tools in its annexation of Crimea.[20] With the annexation essentially complete, in March 2014 President Putin claimed Russian sovereignty over Crimea: "Crimea is our common historical legacy and a very important factor in regional stability. And this strategic territory should be part of a strong and stable sovereignty, which today can only be Russian."[21] Russia's commitment to the priority of the principle of state sovereignty—at least within its own sphere of influence—extends beyond its own borders and immediate neighbours. In 2019 Putin confirmed that "Russia will continue to support Syria's government and people in their efforts to protect the country's sovereignty and territorial integrity."[22]

In the east, China has claimed sovereignty over a large area of the South China Sea, prompting the Philippines to take China to an international tribunal for breaching the United Nations Convention on the Law of the Sea (UNCLOS).[23] The claim was rejected by the tribunal, which found against China,[24] who immediately rejected the international legal finding on the basis it was protecting its "sovereignty and maritime rights."[25] For China, this ambitious sovereignty claim extends "over the relevant waters as well as the seabed and subsoil thereof,"[26] which, if enforced, would radically alter the balance of maritime power in the region with trade, economic and military implications. It would also reshape the potential uses of cyber

---

20 Michael Kofman et al., "Lessons from Russia's Operations in Crimea and Eastern Ukraine'"(Santa Monica: RAND Corporation, 9 May 2017), https://www.rand.org/pubs/research_reports/RR1498.html.
21 Vladimir Putin, "Address by President of the Russian Federation," President of Russia, 18 March 2014, http://en.kremlin.ru/events/president/news/20603.
22 Vladimir Putin, "Russia to Continue Assisting Syria in Protecting Sovereignty - Putin," TASS Russian News Agency, 21 July 2019, https://tass.com/politics/1069649.
23 United Nations, "United Nations Convention on the Law of the Sea" (United Nations, 1982), https://www.un.org/depts/los/convention_agreements/texts/unclos/unclos_e.pdf.
24 Jeremy Page, "Tribunal Rejects Beijing's Claims to South China Sea," *Wall Street Journal*, 12 July 2016, sec. World, https://www.wsj.com/articles/chinas-claim-to-most-of-south-china-sea-has-no-legal-basis-court-says-1468315137.
25 Tom Phillips, Oliver Holmes, and Owen Bowcott, "Beijing Rejects Tribunal's Ruling in South China Sea Case',"*The Guardian*, 12 July 2016, sec. World news, https://www.theguardian.com/world/2016/jul/12/philippines-wins-south-china-sea-case-against-china; People's Daily, "People's Daily Commented on the South China Sea Arbitration: We Belong to Our Territory',"12 July 2016, http://rmrbimg2.people.cn/data/rmrbwap/2016/07/12/cms_1745920168805376.html.
26 Secretary of Defense, "Annual Report to Congress: Military and Security Developments Involving the People's Republic of China 1029" (Washington D.C.: US Department of Defense, 2 May 2019), 7, Annual Report to Congress: Military and Security Developments Involving the People's Republic of China 1029.

in the area. According to the Tallinn Manual 2.0 on the International Law Applicable to Cyber Operations, "States enjoy sovereignty over any cyber infrastructure located on their territory and activities associated with that infrastructure."[27] It goes on, "the territorial nature of sovereignty also places restrictions on other States' cyber operations directed at cyber infrastructure located in sovereign territory."[28] In other words, if cyber surveillance is a bit like looking over a neighbour's wall, China is attempting to significantly extend the geographical area it controls, thereby forcing its neighbours and competitors to peer over its extended 'virtual' wall from a greater distance.

On their own, China's actions in the South China Sea have been enough to prompt a strong American political and military response to maintain established maritime Freedom of Navigation principles.[29] However, when considered alongside China's actions elsewhere, important links emerge between sovereignty, state rights and strategic control—or at least global influence—over the cyber domain become apparent. China has been consistently seeking to change global cyber governance over the past decade and seeks to subsume "Internet sovereignty" within its broader concept of state sovereignty.[30] In so doing, China is seeking to align its politically and ideologically-driven national security and (mis)information aims with "economic, cultural, social, ecological, national defense, and other domains."[31] Ultimately, China seeks to shift the world away from an open Internet to one that is much more closely controlled and monitored by governments—especially its own.[32] Its cyber capability—for surveillance and other military and civilian applications—appears to be on a relentless march of progress, with international law seemingly powerless to slow or halt China's goals.

---

27 Michael N. (Ed.) Schmitt, *Tallinn Manual 2.0 on the International Law Applicable to Cyber Operations* (Cambridge: Cambridge University Press, 2017), 11.
28 Schmitt, 11.
29 Sam LaGrone, "USS Bunker Hill Conducts 2nd South China Sea Freedom of Navigation Operation This Week," USNI News, 29 April 2020, https://news.usni.org/2020/04/29/uss-bunker-hill-conducts-2nd-south-china-sea-freedom-of-navigation-operation-this-week.
30 Scott J. Shackelford et al., "Spotlight on Cyber V: Back to the Future of Internet Governance?", *Georgetown Journal of International Affairs*, 25 June 2015, https://www.georgetownjournalofinternationalaffairs.org/online-edition/back-to-the-future-of-internet-governance.
31 Elsa Kania et al., "China's Strategic Thinking on Building Power in Cyberspace,", New America, 25 September 2017, http://newamerica.org/cybersecurity-initiative/blog/chinas-strategic-thinking-building-power-cyberspace/.
32 Samm Sacks, "Beijing Wants to Rewrite the Rules of the Internet," *The Atlantic*, 18 June 2018, https://www.theatlantic.com/international/archive/2018/06/zte-huawei-china-trump-trade-cyber/563033/.

During President Trump's time in office, the U.S. spent several years re-orienting the emphasis of its policies from internationalist to nationalist: re-asserting its sovereignty, withdrawing from treaties and conventions, and ensuring it answers to no other legal entity. The 2020 White House Report, *United States Strategic Approach to the People's Republic of China*, sets out not only a "whole-of-government strategy with respect to the People's Republic of China (PRC)," it sets out the basis—the ethic—on which strategy is set.[33] The strategy paper concludes that the Unites States government is "guided by a return *to principled realism*, as articulated by the NSS [National Security Strategy]," and will "continue to protect American interests and advance American influence."[34] In a speech at the UN, President Trump set out his core realist principle in direct terms: "For decades the same tired voices proposed the same failed solutions, pursuing global ambitions at the expense of their own people. But only when you take care of your own citizens will you find a true basis for co-operation…I have put America first, just as you should put your countries first."[35]

Given the extent of cyber and other threats posed by China and Russia, there would appear to be little scope for a drastic reversal of approaches by any subsequent U.S. President. In the first month of his Administration, this was confirmed by President Biden who indicated the U.S. would provide a strong response to attacks or external interference—including cyber—in American life and democracy. He warned: "the days of the United States rolling over in the face of Russia's aggressive actions—interfering with our elections, cyberattacks, poisoning its citizens—are over. We will not hesitate to raise the cost on Russia and defend our vital interests and our people."[36]

If there is an ethic underpinning these actions by Russia, China and the U.S. it is rooted in political realism and state self-interest. Just war ethics do not appear relevant, for the most part, because these states are not at war over these particular actions. Furthermore, none were deemed serious

---

33 The White House, "United States Strategic Approach to the People's Republic of China" (Washington D.C.: The White House, 20 May 2020), https://www.whitehouse.gov/wp-content/uploads/2020/05/U.S.-Strategic-Approach-to-The-Peoples-Republic-of-China-Report-5.24v1.pdf.
34 The White House (Italics added for emphasis).
35 Donald Trump, "Remarks by President Trump to the 75th Session of the United Nations General Assembly", The White House, 22 September 2020, https://trumpwhitehouse.archives.gov/briefings-statements/remarks-president-trump-75th-session-united-nations-general-assembly/.
36 Joseph Anthony Biden, "Remarks by President Biden on America's Place in the World," The White House, 5 February 2021, https://www.whitehouse.gov/briefing-room/speeches-remarks/2021/02/04/remarks-by-president-biden-on-americas-place-in-the-world/.

enough even by regional neighbours, to be considered a *casus belli* cause for war. Moving on from just war ethics to other modes of moral analysis, a deontological, rule-based ethical analysis meets several obstacles. These include: the different sovereignty claims are a means of circumventing the norms of international law, rather than an appeal to law; and if there is any sense of obligation at work, it is to domestic populations or national interests, not to the global community and the wider benefit of humanity. Comparatively, a utilitarian ethic, which is assessed in terms of "ends and consequences, their contribution to human happiness, and the prevention of human suffering,"[37] is similarly inadequate. Realist state self-interest is not concerned with all-of-humanity happiness of the prevention of suffering, it is interested only in the wellbeing of the resident population or, just as likely, the wellbeing of ruling administrations.

Our discussion will now examine another grey, ambiguous area between war and peace, by exploring the ethical implications of cyber surveillance in pre-emptive *versus* preventive military contexts.

Ethics of Pre-emptive and Preventive Cyber Surveillance

Pre-emptive and preventive military action are distinguished principally by the timescales involved, followed by the degree of certainty about an impending attack. Brunstetter and Braun distinguish between responding to "[i]mminent threats of terrorist attack" and preemptive strikes to stop states from acquiring WMD [weapons of mass destruction]."[38] The key question they ask is whether "the threat is looming instead of imminent?"[39] If a threat is imminent—the potential aggressor has the means of attack and has demonstrated an intention to do so—a pre-emptive self-defence strike is typically justified within the just war tradition. Earlier tolerance of preventive strikes has eroded.

In the sixteenth-century, Gentili argued for what is more accurately called preventive action—anticipating a future threat rather than an immediate one: "[W]e ought not to wait for violence to be offered us, if it is safer to meet it halfway…No one ought to wait to be struck, unless he is a fool. One ought to provide not only against an offense which is being

---

37 Norman, *The Moral Philosophers: An Introduction to Ethics*, 92.
38 Daniel Brunstetter and Megan Braun, "From Jus Ad Bellum to Jus Ad Vim: Recalibrating Our Understanding of the Moral Use of Force," *Ethics & International Affairs* 27, no. 1 (2013): 96.
39 Brunstetter and Braun, 100.

committed, but also against one which may possibly be committed."[40] The development of international law and developments in the just war tradition of war ethics in subsequent centuries introduced greater time constraints. Walzer argued in the late twentieth and early twenty-first century that, "Both individuals and states can rightly defend themselves against violence that is imminent but not actual; they can fire the first shots if they know themselves about to be attacked."[41] However, Walzer also acknowledged a "spectrum of anticipation," with a clear, immediate and overwhelming threat at one end, and a future, distant danger that has not yet fully manifested at the other.[42]

Walzer argued against the preventive justification of the 2003 Iraq War.[43] Although his position shifted somewhat over time, the furthest he was prepared to go in the ethical justification of preventive intervention was "force-short-of-war," and preferably no further than "politics-short-of-force."[44] This, in a conflict which President Bush sought to justify on a preventive basis, nine months before hostilities commenced: "We must take the battle to the enemy, disrupt his plans, and confront the worst threats *before they emerge.*"[45] While the 2003 Iraq War might appear far-off from the current and future challenges of cyber surveillance, they are connected by the fundamental need for effective surveillance and accurate, reliable and timely intelligence information on which to make political judgements—with their associated ethical dimensions—about resorting to war, or not.

Dipert is sceptical about the applicability of "widely discussed principles of Just War Theory...to cyberwarfare."[46] One reason for his view is cyber attacks—let alone cyber surveillance—"will not be lethal and will not even result in permanent damage to physical objects."[47] Dipert's emphasis on physical damage and Just War *Theory*[48] prompts two responses. The first is that just war is not, historically speaking, a single theory. It is a

---

[40] Alberico Gentili, "De Iure Belli Libri Tres Chapter XIV: Of Defence on Grounds of Utility," in *The Ethics of War: Classic and Contemporary Readings* (Oxford: Blackwell Publishing, 2006), 376.
[41] Michael Walzer, *Just and Unjust Wars*, Third Edn (New York: Basic Books, 2000), 74.
[42] Walzer, 75.
[43] Peter Lee, *Blair's Just War: Iraq and the Illusion of Morality* (Basingstoke: Palgrave, 2012), 57–65.
[44] Michael Walzer, *Just and Unjust Wars*, Fourth Edn (New York: Basic Books, 2006), xvii.
[45] George W. Bush, "President Bush Delivers Graduation Speech at West Point," The White House, 1 June 2002, https://georgewbush-whitehouse.archives.gov/news/releases/2002/06/20020601-3.html (Italics added for emphasis).
[46] Randall R. Dipert, "The Ethics of Cyberwarfare," *Journal of Military Ethics* 9, no. 4 (16 December 2010): 384.
[47] Dipert, 385.
[48] Emphasis added.

broad Western tradition of thought which has adapted over two millennia in response to technological—such as cyber in this instance—political, social, cultural, philosophical and military developments related to war.[49] Further, just war-based *ad bellum* judgements are made in times of peace to assess whether escalation towards war is justified or not. Second, from Beard's perspective, armed attack and physical harm are not necessary for international aggression; the crucial factor is violation of "the crucial state rights of territorial integrity and/or political sovereignty that constitutes aggression in both international law and Walzer's [just war theory]."[50] However, cyber intrusion does not have the same immediate and physically obvious impact that, say, a military drone missile or bomb strike typically has. Cyber intrusion therefore will not obviously register as aggression. Rid offers an alternative view which is relevant to cyber surveillance, noting the purpose of espionage "is not achieving a goal but to gather the information that may be used to design more concrete instruments or policies."[51]

Amidst these contrasting perspectives, military cyber surveillance appears to have two direct links to just war ethical assessment: direct and indirect. Direct relevance exists where the surveillance breaches the territory and sovereignty of another state in what may or may not be an act of war or application of force-short-of-war. Indirect relevance emerges when cyber surveillance provides the means by which assessments about a prospective enemy's actions and intentions can be made.

Barrett sets out the case that "[l]egally, espionage is neither a just cause nor a war crime" and should not be considered a *casus belli*.[52] But even if cyber surveillance—part of cyber espionage—is not, in itself, a cause for war, such actions still breach state sovereignty, which the Tallinn Manual describes as "a foundational principle of international law."[53] In 2015 a meeting of the United Nations (UN) Group of Governmental Experts (GGE) agreed that, "State sovereignty and international norms and principles that flow from sovereignty apply to the conduct by States of [Information and communications technology (ICT)]-related activities and to their jurisdiction

---

49 James Turner Johnson, *Just War Tradition and the Restraint of War* (Princeton: Princeton University Press, 1981); Peter Lee, "A Genealogy of the Ethical Subject in the Just War Tradition" (London, King's College London, 2010).
50 Beard, "Just War, Cyberwar and Cyber-Espionage," 112.
51 Rid, "Cyber War Will Not Take Place," 20.
52 Barrett, "WARFARE IN A NEW DOMAIN," 6.
53 Schmitt, *Tallinn Manual 2.0 on the International Law Applicable to Cyber Operations*, 11.

over ICT infrastructure within their territory."[54] The UN GGE report also emphasised that in their use of cyber technologies, "States must observe, among other principles of international law, State sovereignty, sovereign equality, the settlement of disputes by peaceful means and non-intervention in the internal affairs of other States" and "respect and protect human rights and fundamental freedoms."[55] While the legality, or otherwise, of breaching state sovereignty using military cyber surveillance is beyond the scope of this chapter, the ethical aspects will continue to be contested.

When states are targeted using cyber means it raises ethical questions along the peace-war spectrum. Dipert has argued that "cyberwarfare appears to be almost entirely unaddressed by the traditional morality and laws of war," and was sceptical about the applicability of "widely discussed principles of Just War Theory" because of a lack of lethality and "permanent damage to physical objects."[56] The nature of cyber surveillance of military or military-related personnel or infrastructure introduces new factors into considerations of pre-emptive and preventive action by a state. However, an additional, crucial element in making an assessment of the ethics of any specific action is knowing what was done, when, and by whom and against whom. In addition, a cyber intrusion may have been conducted via either unwitting or willing third party states. So the ethical calculus about pre-emption and prevention in cyber surveillance is inherently linked to what is known as the attribution problem, the degree of harm inflicted and, consequently, proportionate and targeted responses.

Cyber Surveillance: Ethical Assessment and Political Response

The problem with attribution of cyber operations is well known.[57] When it comes to the cyber domain, this politics of truth plays a crucial role

---

54 UN GGE, "Report of the Group of Governmental Experts on Developments in the Field of Information and Telecommunications in the Context of International Security" (New York: United Nations, 22 July 2015), 12, https://undocs.org/A/70/174.
55 UN GGE, 12.
56 Randall R. Dipert, "The Ethics of Cyberwarfare," *Journal of Military Ethics* 9, no. 4 (16 December 2010): 405, 384-5.
57 Thomas Rid and Ben Buchanan, "Attributing Cyber Attacks," *Journal of Strategic Studies* 38, no. 1–2 (2 January 2015): 4–37, https://doi.org/10.1080/01402390.2014.977382; Florian J. Egloff, "Contested Public Attributions of Cyber Incidents and the Role of Academia," *Contemporary Security Policy* 41, no. 1 (2 January 2020): 55–81, https://doi.org/10.1080/13523260.2019.1677324; Barrett, "WARFARE IN A NEW DOMAIN."

alongside technical and operational analysis when it comes to attribution of cyber surveillance, intrusions and attacks.

If cyber surveillance can be correctly attributed, two questions emerge: how should the ethics of surveillance be assessed? And what are appropriate responses to foreign surveillance? Surveillance, including cyber surveillance, of one state by another exists on a spectrum of behaviour which spans peacetime, escalation of tensions, and warfighting. Complicating matters further, non-state actors may attempt cyber surveillance against states. However, in either case—state/ state activities and state/ non-state activities—cyber surveillance alone is unlikely to reach the *ad bellum* just cause (for war) threshold. Just as attribution of cyber intrusion is difficult to assess, so is the intention of any perpetrator. Consequently, appropriate response requires a political and ethical judgement. Consider the words of President Biden, who identified two types of cyber intrusions by Russia—"interfering with our elections, cyberattacks"—which he would consider "aggressive."[58] Potential U.S. responses were not made explicit but would "raise the cost on Russia," thereby allowing flexibility in the degree of future retaliation.[59]

From a realist perspective, states are obliged to carry out activities that it considers reasonable and proportionate to protect their interests; in the case of cyber surveillance, by trying to understand and anticipate the actions and intentions of an opponent—or ally. A simple rule-based analysis of military cyber surveillance during peacetime might conclude that if law or convention has been broken, then an action is therefore unethical. In contrast, a consequence-based ethical analysis might argue that proportionate military cyber surveillance is necessary to *prevent*, or at least anticipate, potential long-term harms.

Brunstetter and Braun make the case for a re-evaluation of classical just war on the basis that it does not sit easily in the current international political system or respond helpfully to military and security challenges. They propose a theory of *jus ad vim* (just use of force) to help "evaluate the spectrum of options available to statesmen, which range from nonviolence, to force-short-of-war, and ultimately to war itself."[60] The criteria they set out—just cause, last resort, proportionality, right intention, legitimate authority—uses conventional just war vocabulary,

---

58 Biden, "Remarks by President Biden on America's Place in the World."
59 Biden.
60 Brunstetter and Braun, "From *Jus Ad Bellum* to *Jus Ad Vim*: Recalibrating Our Understanding of the Moral Use of Force," 88.

with the addition of probability of escalation, and are applied to uncertain, disputed situations short of war where international law might not be obviously breached.[61] Brooks offers an observation about disputed drone use in the international domain which can be extended to military cyber surveillance: "Up to a point, legal vagueness and ambiguity give states face-saving ways to avoid conflict, enabling them to 'look the other way' if a particular state occasionally engages in challenging but not manifestly illegal behavior."[62]

Bringing these various elements together, judgements about cyber surveillance need to be made in ambiguous, disputed, uncertain situations where attribution is difficult, intentions are unclear, and any over-reaction could result in escalation to potential lethal military force. Once a judgement has been made about a particular cyber incident a second judgement must be made—how to respond. The *jus ad bellum* ethics and *jus ad vim* ethics described above both advocate proportionate responses. It would therefore be easy, but restrictive, to infer some kind of like-for-like action: for example, military cyber surveillance prompting a military cyber surveillance response. But neither statecraft nor ethics is so simple. For example, to allow itself leeway in any its response to cyber incidents, the U.S. has made it clear that "under some circumstances, a disruptive activity in cyberspace could constitute an armed attack."[63] By drawing an equivalence between "disruptive activity in cyberspace" and "armed attack," the U.S. is setting out a framework in which responses to aggressive cyber activities will not be limited to the cyber domain. Consider some of the ethical issues and political responses in the following cyber examples which have either an explicitly military focus or potential military implications.

In June 2020 Prime Minister Scott Morrison announced that Australia was "being targeted by a sophisticated state-based cyber actor" and that the cyber intrusions spanned "a range of sectors, including all levels of government, industry, political organisations, education, health, essential service providers, and operators of other critical infrastructure."[64] He did

---

61 Brunstetter and Braun, "From *Jus Ad Bellum* to *Jus Ad Vim*: Recalibrating Our Understanding of the Moral Use of Force."
62 Brooks, "Drones and the International Rule of Law," 84.
63 UN Secretary-General, "Developments in the Field of Information and Telecommunications in the Context of International Security" (New York: United Nations, 15 July 2011), 18, https://undocs.org/A/66/152.
64 Scott Morrison, "Statement on Malicious Cyber Activity against Australian Networks | Prime Minister of Australia'," Prime Minister of Australia, 19 June 2020, https://www.pm.gov.au/media/statement-malicious-cyber-activity-against-australian-networks.

not name China as the state actor in question, acknowledging that "the threshold for public attribution on a technical level is extremely high."[65] It was Peter Jennings of the Australian Strategic Policy Institute (ASPI) who made the public accusation that China was responsible,[66] thereby avoiding the need for China to respond directly against the Australian government. Thereafter, claim and counter-claim ensued, with the Chinese Ministry of Foreign Affairs spokesman Zhao Lijian rejecting the claims as "completely baseless"[67] and attempting to discredit the ASPI by linking their funding to the Australian Defence industry.[68]

Both Australia and China competed to create and sustain their own "truth" about the events in question. If it is the case that China was trying to understand the "political intent and political thinking" of Australian leaders[69]—and only that—then from a political realist perspective, China was looking out for its own interests as part of common international relations practice. However, if the intrusions involved the theft of intellectual property then the harms—actual and potential—would justify a more robust response from the Australian government, beginning with a public naming and shaming: even where China was not actually named.

In the cyber domain where elaborate hoaxes, phishing, spoofing and other deceptions are standard practice, the ethical criterion of intention—from either *jus in bello* or *jus ad vim*—is particularly difficult to establish. Widespread, enduring cyber intrusions against a range of Australian governmental and non-governmental organisations might have had only one aim, or many. The true focus could be one military or defence-related programme, with everything else providing a distraction. Understanding the speed and nature of response could be the "real" purpose of the cyber intrusions. It would be the cyber equivalent of regular Russian assessments

---

[65] Bernard Lagan, "Cyberattack on Australia 'Bears Hallmark of Beijing,'" *The Times*, 20 June 2020, sec. world, https://www.thetimes.co.uk/article/china-believed-to-be-behind-australian-cyberattack-gz52j23s7.
[66] ABC News, "China Cyber Attack Claims Are "Laughable Nonsense," Australian Strategic Policy Institute Says', ABC News, 20 June 2020, https://www.abc.net.au/news/2020-06-20/aspi-denies-sole-org-blaming-china-for-hacking/12376478.
[67] ABC News.
[68] Businessfast News, "China Cyber Attack Claims Are 'Laughable Nonsense,'" Australian Strategic Policy Institute Says - ABC News, *Business Fast* (blog), 20 June 2020, https://www.businessfast.co.uk/china-cyber-attack-claims-are-laughable-nonsense-australian-strategic-policy-institute-says-abc-news/.
[69] ABC News, "China Cyber Attack Claims Are "Laughable Nonsense," Australian Strategic Policy Institute Says'.

of UK and NATO reaction times and methods when intercepting Russian bombers bound for UK airspace.⁷⁰

Accusations made against China by other countries in 2020 support the Australian position in the contest to establish the "truth" of the matter. Also in 2020, the U.S. government identified numerous examples of China's illicit cyber activity, including: conducting or supporting "unauthorized cyber intrusions into United States companies' networks to access sensitive information and trade secrets," for the purpose of "cyber and other types of espionage and theft."⁷¹ Previously, in 2019, the U.S. recognised that limiting responses to "legal and diplomatic instruments of power" was insufficient to deter cyber-enabled theft of American military and other intellectual property (IP).⁷² It was reported that the U.S. started "conducting counter-cyberattacks against Chinese intelligence and military targets."⁷³ A threshold had been crossed: ethically, legally and politically.

Summary

If published estimates of China conducting $200-$600 billion of IP theft from the U.S. each year—with an emphasis on military technology—are accurate, some form of counter-cyber attack would, ethically, seem to be a proportionate response. Preventing or limiting cyber surveillance, theft and other activities—military or otherwise —might be more effective and less costly over time. Such prevention, however, is not just a technical or operational challenge. The U.S. has acted loudly and overtly at the geopolitical level to oppose China's burgeoning surveillance and other cyber capabilities. In September 2020 U.S. Undersecretary for Economic Affairs visited several European capitals to pressure governments—NATO allies—over their choice of future fifth-generation (5G) mobile communications networks. Specifically, he warned against the use of the company Huawei, describing it as "an arm of the CCP [Chinese Communist

---

70 BBC, "Fife-Based RAF Jets Scrambled to Supersonic Bombers," *BBC News*, 15 September 2020, sec. Scotland, https://www.bbc.co.uk/news/uk-scotland-54162051.
71 The White House, "United States Strategic Approach to the People's Republic of China." 3, 9.
72 Erica D. Borghard, "Chinese Hackers Are Stealing U.S. Defense Secrets: Here Is How to Stop Them," Council on Foreign Relations, 11 March 2019, https://www.cfr.org/blog/chinese-hackers-are-stealing-us-defense-secrets-here-how-stop-them.
73 Bill Gertz, "U.S. Hits Back against Chinese Cyberattacks," *The Washington Times*, 6 March 2019, https://www.washingtontimes.com/news/2019/mar/6/us-counters-china-cyberattacks/.

Party] surveillance state and a tool for human rights abuse."[74] The range of responses to military cyber surveillance has expanded to include, legal, diplomatic, direct counter-cyber activities, and economic elements. With geopolitical and economic rivalry escalating among major powers, military cyber surveillance will become increasingly important in the pursuit of national defence and economic self-interest. Appropriate ethical assessment offers political leaders a way of identifying proportionate responses in unpredictable and potentially volatile times.

References

ABC News. "China Cyber Attack Claims Are "Laughable Nonsense", Australian Strategic Policy Institute Says". ABC News, 20 June 2020. https://www.abc.net.au/news/2020-06-20/aspi-denies-sole-org-blaming-china-for-hacking/12376478.
Barrett, Edward T. "Warfare in a New Domain: The Ethics of Military Cyber-Operations". *Journal of Military Ethics* 12, no. 1 (April 2013): 4–17. https://doi.org/10.1080/15027570.2013.782633.
Bartelson, Jens. *A Genealogy of Sovereignty*. Cambridge: Cambridge University Press, 1995.
BBC. "Fife-Based RAF Jets Scrambled to Supersonic Bombers". *BBC News*, 15 September 2020, sec. Scotland. https://www.bbc.co.uk/news/uk-scotland-54162051.
Beard, Matthew. "Just War, Cyberwar and Cyber-Espionage". In *Ethics and the Future of Spying*, 107–19. Oxford: Routledge, 2016.
Biden, Joseph Anthony. "Remarks by President Biden on America"s Place in the World". The White House, 5 February 2021. https://www.whitehouse.gov/briefing-room/speeches-remarks/2021/02/04/remarks-by-president-biden-on-americas-place-in-the-world/.
Borghard, Erica D. "Chinese Hackers Are Stealing U.S. Defense Secrets: Here Is How to Stop Them". Council on Foreign Relations, 11 March 2019. https://www.cfr.org/blog/chinese-hackers-are-stealing-us-defense-secrets-here-how-stop-them.
Brooks, Rosa. "Drones and the International Rule of Law". *Journal of Ethics and International Affairs* 28 (2014): 83–103.
Brunstetter, Daniel, and Megan Braun. "From Jus Ad Bellum to Jus Ad Vim: Recalibrating Our Understanding of the Moral Use of Force". *Ethics & International Affairs* 27, no. 1 (2013): 87–106.

---

[74] Robin Emmott, "U.S. Renews Pressure on Europe to Ditch Huawei in New Networks," *Reuters*, 29 September 2020, https://www.reuters.com/article/us-usa-huawei-tech-europe-idUKKBN26K2MY.

Bush, George W. "President Bush Delivers Graduation Speech at West Point". The White House, 1 June 2002. https://georgewbush-whitehouse.archives.gov/news/releases/2002/06/20020601-3.html.

Businessfast News. "China Cyber Attack Claims Are "Laughable Nonsense", Australian Strategic Policy Institute Says - ABC News". *Business Fast* (blog), 20 June 2020. https://www.businessfast.co.uk/china-cyber-attack-claims-are-laughable-nonsense-australian-strategic-policy-institute-says-abc-news/.

Dipert, Randall R. "The Ethics of Cyberwarfare". *Journal of Military Ethics* 9, no. 4 (16 December 2010): 384–410.

Egloff, Florian J. "Contested Public Attributions of Cyber Incidents and the Role of Academia". *Contemporary Security Policy* 41, no. 1 (2 January 2020): 55–81. https://doi.org/10.1080/13523260.2019.1677324.

Elshtain, Jean Bethke. *Sovereignty*. New York: Basic Books, 2008.

Emmott, Robin. "U.S. Renews Pressure on Europe to Ditch Huawei in New Networks". *Reuters*, 29 September 2020. https://www.reuters.com/article/us-usa-huawei-tech-europe-idUKKBN26K2MY.

Gentili, Alberico. "De Iure Belli Libri Tres Chapter XIV: Of Defence on Grounds of Utility." In *The Ethics of War: Classic and Contemporary Readings*, 271–377. Oxford: Blackwell Publishing, 2006.

Gertz, Bill. "U.S. Hits Back against Chinese Cyberattacks." *The Washington Times*, 6 March 2019. https://www.washingtontimes.com/news/2019/mar/6/us-counters-china-cyberattacks/.

Jinghua, Lyu. "What Are China"s Cyber Capabilities and Intentions?" Carnegie Endowment for International Peace, 1 April 2019. https://carnegieendowment.org/2019/04/01/what-are-china-s-cyber-capabilities-and-intentions-pub-78734.

Johnson, James Turner. *Just War Tradition and the Restraint of War*. Princeton: Princeton University Press, 1981.

Kania, Elsa, Samm Sacks, Paul Triolo, and Graham Webster. "China's Strategic Thinking on Building Power in Cyberspace." New America, 25 September 2017. http://newamerica.org/cybersecurity-initiative/blog/chinas-strategic-thinking-building-power-cyberspace/.

Kaspersky. "Kaspersky Researchers Uncover New Targeted Campaign against Financial and Military Organizations | Kaspersky." Kaspersky, 13 August 2020. https://www.kaspersky.com/about/press-releases/2020_kaspersky-researchers-uncover-new-targeted-campaign-against-financial-and-military-organizations.

Kofman, Michael, Katya Migacheva, Brian Nichiporuk, Andrew Radin, Olesya Tkacheva, and Jenny Oberholtzer. "Lessons from Russia's Operations in Crimea and Eastern Ukraine." Santa Monica: RAND Corporation, 9 May 2017. https://www.rand.org/pubs/research_reports/RR1498.html.

Lagan, Bernard. "Cyberattack on Australia 'Bears Hallmark of Beijing.'" *The Times*. 20 June 2020, sec. world. https://www.thetimes.co.uk/article/china-believed-to-be-behind-australian-cyberattack-gz52j23s7.

LaGrone, Sam. "USS Bunker Hill Conducts 2nd South China Sea Freedom of Navigation Operation This Week." USNI News, 29 April 2020. https://news.

usni.org/2020/04/29/uss-bunker-hill-conducts-2nd-south-china-sea-freedom-of-navigation-operation-this-week.

Lee, Peter. "A Genealogy of the Ethical Subject in the Just War Tradition." King"s College London, 2010.

———. *Blair's Just War: Iraq and the Illusion of Morality.* Basingstoke: Palgrave, 2012.

Morrison, Scott. "Statement on Malicious Cyber Activity against Australian Networks | Prime Minister of Australia." Prime Minister of Australia, 19 June 2020. https://www.pm.gov.au/media/statement-malicious-cyber-activity-against-australian-networks.

Norman, Richard. *The Moral Philosophers: An Introduction to Ethics.* Oxford: Oxford University Press, 1998.

Paganini, Pierluigi. "Chinese APT CactusPete Targets Military and Financial Orgs in Eastern Europe." Security Affairs, 14 August 2020. https://securityaffairs.co/wordpress/107128/apt/catuspete-updated-backdoor.html.

Page, Jeremy. "Tribunal Rejects Beijing's Claims to South China Sea." *Wall Street Journal*, 12 July 2016, sec. World. https://www.wsj.com/articles/chinas-claim-to-most-of-south-china-sea-has-no-legal-basis-court-says-1468315137.

People"s Daily. "People's Daily Commented on the South China Sea Arbitration: We Belong to Our Territory." 12 July 2016. http://rmrbimg2.people.cn/data/rmrbwap/2016/07/12/cms_1745920168805376.html.

Phillips, Tom, Oliver Holmes, and Owen Bowcott. "Beijing Rejects Tribunal's Ruling in South China Sea Case." *The Guardian*, 12 July 2016, sec. World news. https://www.theguardian.com/world/2016/jul/12/philippines-wins-south-china-sea-case-against-china.

Putin, Vladimir. "Address by President of the Russian Federation." President of Russia, 18 March 2014. http://en.kremlin.ru/events/president/news/20603.

———. "Russia to Continue Assisting Syria in Protecting Sovereignty - Putin." TASS Russian News Agency, 21 July 2019. https://tass.com/politics/1069649.

Rid, Thomas. "Cyber War Will Not Take Place." *Journal of Strategic Studies* 35, no. 1 (1 February 2012): 5–32. https://doi.org/10.1080/01402390.2011.608939.

Rid, Thomas, and Ben Buchanan. "Attributing Cyber Attacks." *Journal of Strategic Studies* 38, no. 1–2 (2 January 2015): 4–37. https://doi.org/10.1080/01402390.20 14.977382.

Sacks, Samm. "Beijing Wants to Rewrite the Rules of the Internet." The Atlantic, 18 June 2018. https://www.theatlantic.com/international/archive/2018/06/zte-huawei-china-trump-trade-cyber/563033/.

Schmitt, Michael N. (Ed.). *Tallinn Manual 2.0 on the International Law Applicable to Cyber Operations.* Cambridge: Cambridge University Press, 2017.

Secretary of Defense. "Annual Report to Congress: Military and Security Developments Involving the People's Republic of China 1029." Washington D.C.: US Department of Defense, 2 May 2019. Annual Report to Congress: Military and Security Developments Involving the People"s Republic of China 1029.

Shackelford, Scott J., Enrique Oti, Jaclyn A. Kerr, Elaine Korzak, and Andreas Kuehn. "Spotlight on Cyber V: Back to the Future of Internet Governance?" Georgetown Journal of International Affairs, 25 June 2015. https://

www.georgetownjournalofinternationalaffairs.org/online-edition/back-to-the-future-of-internet-governance.

The White House. "United States Strategic Approach to the People's Republic of China." Washington D.C.: The White House, 20 May 2020. https://www.whitehouse.gov/wp-content/uploads/2020/05/U.S.-Strategic-Approach-to-The-Peoples-Republic-of-China-Report-5.24v1.pdf.

Trump, Donald. "National Cyber Strategy of the United States of America." Washington D.C.: The White House, September 2018. https://www.whitehouse.gov/wp-content/uploads/2018/09/National-Cyber-Strategy.pdf.

———. "President Donald J. Trump's Foreign Policy Puts America First." The White House, 30 January 2018. https://www.whitehouse.gov/briefings-statements/president-donald-j-trumps-foreign-policy-puts-america-first/.

UN GGE. "Report of the Group of Governmental Experts on Developments in the Field of Information and Telecommunications in the Context of International Security." New York: United Nations, 22 July 2015. https://undocs.org/A/70/174.

UN Secretary-General. "Developments in the Field of Information and Telecommunications in the Context of International Security." New York: United Nations, 15 July 2011. https://undocs.org/A/66/152.

United Nations. "United Nations Convention on the Law of the Sea." United Nations, 1982. https://www.un.org/depts/los/convention_agreements/texts/unclos/unclos_e.pdf.

Walzer, Michael. *Just and Unjust Wars*. Third. New York: Basic Books, 2000.

———. *Just and Unjust Wars*. Fourth Edn. New York: Basic Books, 2006.

Williams, Gareth D. "Piercing the Shield of Sovereignty: An Assessment of the Legal Status of the 'Unwilling or Unable' Test." *University of New South Wales Law Journal* 36, no. 2 (2013): 619–41.

8

ETHICS AND CYBER ENABLED PSYOP[1]

Adam Henschke

Psychological Operations (PSYOP)[2] involve information operations conducted against groups in order to change their "hearts and minds." While PSYOP have long been a staple of military operations, the rise of cyber space has expanded the use of these operations and broadened the potential targets of such operations. This chapter looks at ethical issues of PSYOP, generally, and how these ethical issues play out in a "cyber context."

The chapter starts with a specific example of recent cyber enabled PSYOP (CE PSYOP) that targeted citizens of liberal democracies. It then gives a general introduction to the notion of PSYOP, and looks at how the rise of cyber space, and its integration into civilian life, presents not just an expanded attack surface for PSYOP, but also raises a set of somewhat novel contexts for ethical analysis. This chapter looks at militaries as the key institution of interest and compares the ethics of traditional kinetic military operations with use of CE PSYOP, seeing where CE PSYOP fit with the *jus ad bellum* and *jus in bello* criteria. The chapter then asks if there are any relevant moral differences between targeting foreign adversaries, foreign allies/non-adversaries and one's own citizens. Having engaged in these conceptual and ethical discussions, the chapter briefly concludes

---

[1] I thank CJ O'Connor for her help with this paper, and her and Ned Dobos for many stimulating and useful discussions that helped develop my thinking on this topic. I also acknowledge the following funders for their support, noting that the views in this chapter are my own and do not represent that of the funders at all: Australia Research Council Discovery Grant DP1801103439 "Intelligence and National Security: Ethics, Efficacy And Accountability," Australian Department of Defence Strategic Policy Grant "Countering Foreign Interference And Cyber War Challenges," and the European Research Council Advanced Grant "Global Terrorism and Collective Moral Responsibility: Redesigning Military, Police and Intelligence Institutions in Liberal Democracies."

[2] For this chapter I follow the US convention of referring to psychological operations as PSYOP, for both singular and plural operations.

with a discussion of what these conceptual and ethical discussions mean for liberal democratic states.

Modern PSYOP and Technology

With the political shock of the election of Donald Trump to U.S. president in 2016 people looked for answers to how such an unexpected outcome could have occurred. Publicly flagged by the then-Office of the Director of National Intelligence administration on 7th October 2016,[3] the role of Russian influence on the U.S. presidential election has haunted Trump since his win. In early releases from the Robert Mueller investigation,[4] a Russian group known as the Internet Research Agency (IRA) were identified as coordinating and running a comprehensive campaign to undermine support for Hillary Clinton and, as the 2016 presidential campaign developed, to increasingly support Donald Trump. Given the connections between the IRA and the Russian government, these efforts have all the hallmarks of a modern psychological operation.

In and of itself, this is nothing new. "PSYOP is as old as warfare itself."[5] Arguably, political actors, as well as private actors, have been involved in PSYOP for the entirety of human history. In the biblical story, Gideon engaged in PSYOP against the Midianites by convincing them that an army of 300 was in fact 30,000.[6] However, what the world witnessed in 2016 was potentially something new. The relevant difference here is the use of "cyber technologies" as vectors for these PSYOP. The targets of the IRA's efforts were U.S. citizens, and the IRA leveraged social media in a range of ways to conduct their operation. Moreover, although the operation involved individuals coming from Russia to be physically present in the U.S., key aspects of the operation were conducted remotely, again, using cyber technologies as vectors for the PSYOP.[7]

---

3 James R Clapper and Trey Brown, *Facts and Fears* (Viking, 2019), 352.
4 Robert S. Mueller, "United States of America V. Internet Research Agency," ed. United States Department of Justice (District Of Columbia2018).
5 Peter J Smyczek, "Regulating the Battlefield of the Future: The Legal Limitations on the Conduct of Psychological Operations (Psyop) under Public International Law," *Air Force Law Review* 57 (2005): 214.
6 Garth Jowett and Victoria O'Donnell, *Propaganda and Persuasion*, First ed. (Newbury Park: Sage Publications, 1986), 121.
7 Mueller, "United States of America V. Internet Research Agency."

The U.S. 2016 election is not a one-off case. The 2018 U.S. midterms were also targeted for CE PSYOP, and as of October 2020, while this chapter was being written, there are reports that Russian, Chinese and Iranian are trying to use cyber technologies to influence the 2020 U.S. election.[8] Moreover, these efforts expand beyond the U.S. There are investigations into the role of foreign actors in the 2016 UK Brexit decision. Similarly, European countries like the Netherlands, France, Germany, and Sweden have all had to develop responses to foreign information operations.[9]

As with many issues involving national security and new technologies, we should ask what makes these cases novel, if anything? Is there any particular difference that makes a *moral* difference? If political actors have been using PSYOP for millennia, then why should we be concerned about these current cases? This chapter seeks to understand if there is a difference that makes a moral difference with regard to modern CE PSYOP. The chapter proceeds as follows. It begins with a clarification of what PSYOP are, offering broad and narrow conceptions, and has a focus on the narrower conception. CE PSYOP are then introduced. This clarifies two points. First, it gives a clear sense of what I mean by CE PSYOP. Second, it gives examples of how CE PSYOP operate using features of cyber space. The chapter then moves to the moral considerations, looking at arguments for and against using CE PSYOP. Finally, it looks at the issue of whether a liberal democratic state can justifiably use PSYOP against its own citizens.

What Are PSYOP?

The simplest idea of PSYOP is that they are information operations conducted against targets with the desire to change their beliefs and/or motivations for some political end. In 1945, Paul Linebarger gave a definition of psychological warfare as "the use of propaganda against an enemy, together with such other operational measures of a military, economic or political nature as may be required to supplement propaganda." And in 1951, Harold Laswell made connections between military success and

---

8 William Evanina, "Statement by Ncsc Director William Evanina: Election Threat Update for the American Public," news release, 7 August, 2020, https://www.dni.gov/index.php/newsroom/press-releases/item/2139-statement-by-ncsc-director-william-evanina-election-threat-update-for-the-american-public.
9 Erik Brattberg and Tim Maurer, "Russian Election Interference: Europe's Counter to Fake News and Cyber Attacks," (Carnegie Endowmnent For International Peace, 2018).

efforts to change an enemy's hearts and minds: "The basic idea is that the best success in war is achieved by the destruction of the enemy's will to resist, and with a minimum annihilation of the fighting capacity."[10] More recently, the official U.S. *Psychological Operations Joint Publication 3-13.2* defines psychological operations as:

> [p]lanned operations to convey selected information and indicators to foreign audiences to influence their emotions, motives, objective reasoning, and ultimately the behavior of foreign governments, organizations, groups, and individuals. The purpose of psychological operations is to induce or reinforce foreign attitudes and behavior favorable to the originator's objectives.[11]

The use of information is core to any PSYOP, but information is taken in a broad sense, in that it is not necessarily true, or fact-based. This point is important to note as in accounts of information like that of Luciano Floridi and Fred Dretske, information must be true in order for it to be information.[12] I do not take that position here, and, in line with what I have written on information elsewhere, we can simply take information to mean data that is well ordered, has semantic content and might be *judged to be true*, rather than necessarily being true.[13] I also note that information in PSYOP can include deliberate use of information (broadly construed) to change a target's cognitive environment. For instance, the U.S. military and police use extreme music played at high volumes to wear down a target's psychological states in order to bring about the desired ends.[14]

I do, however, need to make a distinction between different PSYOP. In a broad conception, a bullying campaign against a child at school would count as PSYOP. Those conducting the campaign are conveying "selected information and indicators to [an audience] to influence their emotions,

---

10 Both cited in Jowett and O'Donnell, *Propaganda and Persuasion*, 119.
11 Chairman Of The Joint Chiefs Of Staff, "Psychological Operations: Revision of Joint Publication 3-13.2," (Joint Chiefs Of Staff, 2010), GL-8.
12 Fred Irwin Dretske, *Knowledge and the Flow of Information* (Oxford: B. Blackwell, 1981); Luciano Floridi, "Semantic Conceptions of Information," in *The Stanford Encyclopedia Of Philosophy*, ed. Edward N. Zalta (2011); *The Philosophy of Information* (Oxford: Oxford University Press, 2011).
13 Adam Henschke, *Ethics in an Age of Surveillance: Virtual Identities and Personal Information* (New York: Cambridge University Press, 2017), 134-40.
14 Justin Caba, "Torture Methods with Sound: How Pure Noise Can Be Used to Break You Psychologically," *Medical Daily*, 20 January 2015.

motives, objective reasoning."[15] However, I do not wish to include these sorts of campaigns in the discussion here: In order to mark out what is of interest for this chapter, what is of relevance is that the PSYOP is for some political end, broadly construed—a point in line with the earlier accounts offered by Linebarger and Lerner. This can be a military end, a national security intelligence end, or political end; however, it is not limited to that.

We can also identify a rough taxonomy of PSYOP. Like in the differentiation of propaganda offered by Garth Jowett and Victoria O'Donnell,[16] we can consider that PSYOP could potentially be "white," "black" and "grey." "First, overt messages are called "White propaganda" or "White PSYOP"… White PSYOP are those messages issued from an open and acknowledged source, targeting a specific audience and not hiding the source from the enemy or indeed the world… White messages are truthful in nature and are based on objective fact."[17] White PSYOP are marked by two essential elements. First, the source is deliberately identifiable. Second, the content is—at least in part—accurate or truthful.[18] Think here of a political campaign or advertising. These are PSYOP. However, if the political candidate or party is identified with the campaign, and there are at least aspects of that campaign or advertising that are truthful, or at very least truth verifiable, then we have an example of white PSYOP. For instance, a political campaign that promises to cut taxes to make the voter's lives better is white PSYOP: the political actors are identified with the information campaign, and should the particular candidate be elected, they can be held to account on whether they cut taxes or not. In contrast, in black PSYOP the source is unknown or at least disguised, and the content is either deliberately untruthful, or devoid of any content that can be verified.[19] "Black PSYOP consist of messages from an unknown source…and are often based on lies or fabrications."[20] An uncredited social media campaign that attacks a particular political opponent for being in league with paedophiles

---

15 Chairman Of The Joint Chiefs Of Staff, "Psychological Operations: Revision of Joint Publication 3-13.2," GL-8.
16 I note there that I am using the first edition of *Propaganda and Persuasion* in this chapter, from 1986. The 2018 version is the seventh edition and has been considerably updated since the first edition. However, I have elected to stay with the first edition in recognition that many of the problems of PSYOP are persistent issues.
17 Smyczek, "Regulating the Battlefield of the Future: The Legal Limitations on the Conduct of Psychological Operations (Psyop) under Public International Law," 215.
18 Jowett and O'Donnell, *Propaganda and Persuasion*, 17.
19 *Propaganda and Persuasion*, 18.
20 Smyczek, "Regulating the Battlefield of the Future: The Legal Limitations on the Conduct of Psychological Operations (Psyop) under Public International Law," 218.

would be black PSYOP.[21] Similarly, blasting extreme music at high volume at a target to break them psychologically would be black PSYOP, even if the source were known, as there is no connection between the shaping of the cognitive environment and truth. Grey PSYOP "activities fall between the two extremes and are neither completely true nor completely false… and do not specifically identify the source."[22] In the age of social media, the "source" might be known, but it is hard to glean the actual people behind the operation, and/or "the accuracy of the information is uncertain."[23] The Mueller investigation identified a number of efforts in the U.S. 2016 election to amplify anti-Clinton messaging, where the publicly identifiable sources were not the actual origins of that information.[24] Again, the content is also uncertain—it is either hard to know if it is truthful or the content is something that is mostly devoid of any verifiable content. I would expect that many common political advertising campaigns would be grey PSYOP.

Cyber Enabled PSYOP

CE PSYOP are thus information operations conducted against targets with the desire to change their beliefs and/or motivations for some political end, *which involve cyber technologies in some essential way*. This involvement might use cyber technologies as the means, or as the proximate target. On cyber technologies as the means, the PSYOP use some information or communication technology or combination of technologies as a necessary element of the information operation. On cyber technologies as the target, the technologies themselves are the proximate target of the operation, though in this situation, there is going to be some further ultimate end.

Cyber as the means: A basic example of cyber as the means is the use of social media to change the target's beliefs and/or motivations. This is where *for some political end* becomes essential to the account of PSYOP. Without such a caveat, any social media advertising would be CE PSYOP. The important aspect here is that given that cyber technologies like social

---

[21] This example is chosen specifically to match key aspects of the QAnon conspiracy theory which centres on the idea that elites, including high profile U.S. Democrats are covering up an extensive global child trafficking operation, Andrew Griffin, "What Is Qanon? The Origins of Bizarre Conspiracy Theory Spreading Online," *Independent*, 24 August 2020.
[22] Smyczek, "Regulating the Battlefield of the Future: The Legal Limitations on the Conduct of Psychological Operations (Psyop) under Public International Law," 218.
[23] Jowett and O'Donnell, *Propaganda and Persuasion*, 17.
[24] Mueller, "United States of America V. Internet Research Agency."

media enable mass distribution of information and communications that are largely unbound by geography or time, and are so woven into everyday life, they present a unique PSYOP vector.

Cyber as the target: Another aspect of CE PSYOP involves direct attacks on cyber infrastructure to change adversarial behaviour. Stuxnet, is an infamous cyber weapon that caused physical destruction to centrifuges of the Iranian nuclear program.[25] However, as Thomas Rid has argued, while the proximate target was physical damage to the centrifuges, we can also understand Stuxnet as targeting the *trust* of the Iranian nuclear engineers in their own systems and capacity to develop and operate a nuclear enrichment program.[26] Hence, while the proximate target of the cyber weapon might have been the physical infrastructure, a potential ultimate target was the beliefs and motivations of the Iranian engineers. While cyber as the target of PSYOP poses an interesting area for discussion, it is not the focus here. The remaining discussion of the chapter will be on cyber as the means.

Looking at cyber as the target allows us to recognize the importance of PSYOP being complex actions involving both proximate and ultimate aims.[27] As Michael Robillard has argued, we ought to view terrorism as being a form of PSYOP.[28] This point is important as terrorism necessarily involves violence directed at a proximate target (innocent civilians) in order to bring about some ultimate political, ideological or religious end[29] through public communications.[30] Many complex PSYOP will take a similar structure, and PSYOP that use cyber as the means are no different. To return to our earlier example, the IRA's efforts to exploit social media likely had proximate and ultimate ends. The proximate ends were to amplify opposition to Hillary Clinton in 2016, and to amplify support for Donald Trump towards the

---

25 Ralph Langner, "To Kill a Centrifuge: A Technical Analysis of What Stuxnet's Creators Wanted to Achieve," (Arlington: The Langner Group, 2013).
26 Thomas Rid, *Cyber War Will Not Take Place* (Hurst & Company, 2013), 32-34.. Adding credence to this view, Thomas Rid cites a participant in the attack who claimed that "[t]he intent was that the failures should make them feel they were stupid, which is what happened" *Cyber War Will Not Take Place*, 32.
27 For more on the idea of complex actions, see Seumas Miller's discussion of joint actions Seumas Miller, *The Moral Foundations of Social Institutions: A Philosophical Study* (Cambridge: Cambridge University Press, 2010).
28 Michael Robillard, "Counter-Terrorism Ethics and Psyop," in *Counter-Terrorism: The Ethical Issues* ed. Seumas Miller, Adam Henschke, and Jonas Feltes (Edward Elgar, 2021).
29 Igor Primoratz, "What Is Terrorism?," in *Terrorism: The Philosophical Issues*, ed. Igor Primoratz (Basingstoke: Palgrave, 2004).
30 Seumas Miller and Jonas Feltes, "Defining Terrorism," in *Counter-Terrorism: The Ethical Issues* ed. Seumas Miller, Adam Henschke, and Jonas Feltes (Edward Elgar, 2021).

end of the 2016 campaign[31], and then into 2018[32] and 2020.[33] However, the ultimate ends are to sow discord and distrust in the results of an election outcome. Much like Rid's suggestion about undermining trust being one of the ultimate ends of the Stuxnet operation, the IRA and other's CE PSYOP ultimate ends for information operations in election campaigns may be to undermine the trust in core democratic processes. This degradation in trust can then lead to voter disengagement and demotivation to vote, increased political and social instability in any given outcome, and can also act as the basis for propaganda efforts that seek to discredit democratic processes more generally.

The important thing here to note is how cyber technologies change or evolve PSYOP. Again, PSYOP are nothing new, so what makes CE PSYOP worthy of attention? Cyber technologies are now deeply woven into civilian life. Many people are immersed in unsourced and unverifiable information through technologies like smart phones, and an increasing number of people rely on social media as their primary information source, with social media overtaking print media for U.S. people's primary news source in 2018.[34] This integration has two implications. First, it means that civilian targeting is comparatively cheap and easy—the cost of the IRA operation in 2016 has been estimated at US$ 1.25 million per month in 2016.[35] While that is a significant investment, compared to how much an offensive military operation or even covert military operation would cost, it is relatively cheap and easy. Second, it means that those wishing to target large numbers of civilians can do so from a geographic distance. Thus, the capacity for black and grey PSYOP directed against whole civilian populations is significantly increased through cyber means.

A further more subtle aspect, is that this information is both unverifiable and due to its carriage on social media—may be more trusted by a recipient. By unverifiable, I mean that the origin of the information can be hard to find and is easily obscured. Again, by definition, this makes a significant proportion of information on social media black or grey. The compounding factor is that if a person receives some information from a trusted friend or

---

31 Mueller, "United States of America V. Internet Research Agency."
32 *United States of America V. Elena Alekseevna Khusyaynova*, (2018).
33 Evanina, "Statement by Ncsc Director William Evanina: Election Threat Update for the American Public."
34 Elisa Shearer, "Social Media Outpaces Print Newspapers in the U.S. As a News Source," news release, 10 December, 2018, https://www.pewresearch.org/fact-tank/2018/12/10/social-media-outpaces-print-newspapers-in-the-u-s-as-a-news-source/.
35 Mueller, "United States of America V. Internet Research Agency."

family member, as they are a trusted person, the information is more likely to be trusted.[36] Thus, cyber technologies, specifically information carried by social media, have a greater potential to be effective PSYOP platforms due to the trusted nature of those who share the information, even if the source is unknown and the quality of the information is unverified.

Can Aggressive Cyber Enabled PSYOP Be Justified?

In this section, I look at potential reasons that might justify "aggressive" CE PSYOP, and the arguments against CE PSYOP. As will be shown, context and intent matter. One cannot simply say "CE PSYOP are good" or "CE PSYOP are bad." We need to see the context that they are being used in, and the reason that they are being used. By aggressive CE PSYOP, I mean CE PSYOP activities that are used in offensive or attacking manner against an adversary state and/or its people.

The first and "easiest" way to justify CE PSYOP is to see such an operation as part of a military campaign. The justification is twofold. First, consider "traditional" or kinetic military actions, where people die and things like critical infrastructure essential for civilian life are destroyed. By comparison, PSYOP generally, and CE PSYOP in particular, become a much better option. If you can defeat an enemy without going to war, without putting your own soldiers at risk, and without physically harming the civilians of the target nation, CE PSYOP would seem to be the morally preferred option.

However, as with other ethical issues around cyber warfare, things are not so cut and dried.[37] First, in line with the just war tradition, any military decisions would have to meet the *jus ad bellum* criteria. Comparing CE PSYOP to the death and destruction of kinetic warfare may seem to meet the proportionality criterion (and this is not certain, see below), but they would also have to have a just cause, be done with the right intention, be conducted by a legitimate authority, have a reasonable hope of success and be the last resort. If CE PSYOP are part of a larger traditional military campaign that meets these conditions, then there is a potential that they

---

36 Nicolas M. Anspach, "The New Personal Influence: How Our Facebook Friends Influence the News We Read," *Political Communication* 34, no. 4 (2017).
37 For instance, see *Binary Bullets* for a range of views on the ethics of cyberwarfare Fritz Allhoff, Adam Henschke, and Bradley Jay Strawser, eds., *Binary Bullets: The Ethics of Cyberwarfare* (Oxford University Press, 2016).

would meet the *jus ad bellum* criteria.[38] That said, just cause and right intention pose significant ethical challenges for CE PSYOP. Because they are non-kinetic, they do not need a *casus belli*, so are "easy" to deploy without just cause. The point here is threefold. First, does the possession of CE PSYOP capability lower the threshold on decisions to go war? In addition to issues of last resort (see below), the worry is that the non-kinetic nature of CE PSYOP can seem like such operations can be justified by a lower just cause than traditional military operations. This leads us to the second point; what sort of events are needed for CE PSYOP to be justified?[39] This is particularly important when considering CE PSYOP that are conducted by a state's military. Finally, and parallel to this, CE PSYOP will likely be used in a way to deliberately interfere in the social or political integrity of sovereign nations. In addition to issues of discrimination (see below), do efforts to undermine the social integrity and interfere in a sovereign nation's political processes meet with just cause and right intention? We can, perhaps look to discussions of justifications of intervention and the responsibility to protect.[40] The point here is not to offer any solutions on how just cause and right intention ought to operate with regard to CE PSYOP. Instead, the point is to show that these are complex areas and need to be explored further.

Similarly, the reasonable hope of success criterion is equally complex in this situation. First, are we assessing the reasonable hope of success of this CE PSYOP, or the reasonable hope of success of the overall military campaign? If it is the CE PSYOP alone, then it seems hard to expect that such an operation will be successful. Moreover, what counts as success, and what *should* count as success? Garth Jowett and Victoria O'Donnell talk about how the PSYOP campaigns conducted by the British against the Germans in World War One, might have led to a disinclination for people

---

38 For more on the just war tradition, see: Tony J. Coates, *The Ethics of War* (Manchester University Press, 1997); Stephen Coleman, *Military Ethics: An Introduction with Case Examples* (New York: Oxford University Press, 2013); Jeff McMahan, *Killing in War* (Oxford: Clarendon Press, 2009); Brian Orend, *The Ethics of War*, 2nd ed. (Vancouver: University Of Alberta, 2013); Gregory Reichberg, Henrik Syse, and Endre Begby, eds., *The Ethics of War: Classic and Contemporary Readings* (Blackwell, 2006); Michael Walzer, *Just War and Unjust Wars*, 4th ed. (New York: Basic Books, 2006).

39 This challenge about non-kinetic military operations and just cause has been discussed by Shannon Brandt Ford, "*Jus Ad Vim* and the Just Use of Lethal Force-Short-of-War," in *Routledge Handbook of Ethics and War: Just War in the 21st Century*, ed. Fritz Allhoff, Nicholas G. Evans, and Adam Henschke (Routledge, 2013).

40 See Ned Dobos on insurrection and Ramesh Thakur for more one these issues Ned Dobos, *Insurrection and Intervention: The Two Faces of Sovereignty* (Cambridge University Press, 2011); Ramesh Thakur, "The Responsibility to Protect at 15," *International Affairs* 92, no. 2 (2016).

to believe the threat posed by the Nazis in the lead up to World War Two. "The very success of the British propaganda efforts in 1914-1918 proved to be a serious handicap in getting the world to accept the reality of what was happening in Nazi Germany, and this created a disastrous delay in the public's awareness of the horrors of the concentration camps and other Nazi atrocities."[41] We can see similar risks with modern CE PSYOP, particularly black CE PSYOP—while such disinformation campaigns might be useful in the short term, their long-term success is much harder to ascertain. Consider for instance, that U.S. forces used CE PSYOP to convince Iraqis to withdraw support for so-called IS (Islamic State), by making promises that with so-called IS gone, their lives would be better. However, following the military defeat of so-called IS, the local conditions did not get noticeably better, and five years later local conditions are still horrible.[42] It is conceivable that not only will the locals be suspicious of foreign promises, but having unmet expectations can provide fertile ground for so-called IS, or some other similar group, to claw back local support.

When thinking of the last resort criterion, things again become complex. If CE PSYOP are part of a justified ongoing military campaign, and that campaign was chosen as a last resort, then this would likely be permissible. However, CE PSYOP are likely to be used in advance of a kinetic campaign.[43] So, now CE PSYOP are not the last resort, but would be considered something used prior to the last resort. Rather than being considered under the *jus ad bellum* criteria, this places CE PSYOP in the area of *jus ad vim*, "force-short-of-war,"[44] or perhaps in the realm of the "just intelligence tradition."[45] We do not have the space to cover the detail of *jus ad vim* or the just intelligence tradition here, but these areas are complex, and it is not immediately clear whether CE PSYOP would be ethically preferred to a kinetic military campaign. The point here is that it is hard to see whether CE PSYOP would fit with the last resort criterion, or if they need to be considered by a different set of ethical criteria.

---

41 Jowett and O'Donnell, *Propaganda and Persuasion*, 137.
42 While this is intended to be a fictional account of what might happen, it does roughly track the U.S. experience in Iraq. See, for instance, David Kilcullen's *Blood Year* David Kilcullen, *Blood Year: The Unravelling of Western Counterterrorism* (Oxford University Press, 2016).
43 Robillard, "Counter-Terrorism Ethics and Psyop."
44 For more on *jus ad vim* see: Ford, "*Jus Ad Vim* and the Just Use of Lethal Force-Short-of-War."
45 For more on the "just intelligence tradition" see Ross Bellaby, *The Ethics of Intelligence* (Routledge, 2014); David Omand and Mark Phythian, *Principled Spying: The Ethics of Secret Intelligence* (Oxford University Press, 2018); David L Perry, *Partly Cloudy: Ethics in War, Espionage, Covert Action, and Interrogation* (Rowman & Littlefield, 2016).

This also leads us to the problem of proportionality. On the face of it, if we are selecting between the use of CE means versus using kinetic military force, CE PSYOP would seem to be the proportionate option. Consider that a well-orchestrated and planned CE PSYOP campaign destabilised a particular political leadership, and a "better" set of leaders took their place.[46] This might parallel the just cause and right intention of a traditional military campaign, but avoids the bloodshed and destruction, thus, it seems to be the proportionate choice. However, the CE PSYOP could potentially undermine long term trust and belief in the political processes, could unleash a range of simmering tensions leading to sectarian violence and instability like what happened in Rwanda, and—as Jowett and O'Donnell noted—happened in the interwar years with Germany, so could ultimately lead to people turning a blind eye to evolving threats because they believe the stories about what is happening to be fabrications like the CE PSYOP campaign. The important thing with these negative scenarios is that the cyber technologies mean that the CE PSYOP campaign can spread beyond the target area and population. Unlike in the past, the global scope of information communication technologies (ICTs) means that a targeted CE PSYOP campaign can quickly and rapidly be picked up elsewhere, with potentially disastrous results. Consider the way that ICTs spread the "Muhammed cartoons" beyond their initial publication in Denmark. Further, CE PSYOP can leave a residue of disinformation long after the campaign ends, a phenomenon Emily Thorson has called "Belief Echoes."[47] We need only think of how the initial publication of connections between the MMR (measles, mumps and rubella) vaccine and autism continues to fuel anti-vaxxer beliefs.[48] There is a range of concerns about proportionality here. The long-term consequences of PSYOP and propaganda campaigns are particularly hard to predict. Think here of the persistence of anti-Semitic beliefs tied to the propaganda publication "The Protocols Of The Elders Of Zion." This publication, first published in the early twentieth-century, still gives a foundation to anti-Semitism more than a century later. The point here is that a CE PSYOP campaign does not necessarily meet the proportionality criterion so easily. Further, the long-term consequences of the CE PSYOP are very hard to predict and calculate.

---

46 Overlooking here what "better" means.
47 Emily Thorson, "Belief Echoes: The Persistent Effects of Corrected Misinformation," *Political Communication* 33, no. 3 (2016).
48 Brett Bricker and Jacob Justice, "The Postmodern Medical Paradigm: A Case Study of Anti-Mmr Vaccine Arguments," *Western Journal Of Communication* 83, no. 2 (2019).

Having touched on *jus ad bellum* proportionality, we also need to consider the *jus in bello* criteria of proportionality and discrimination. As before, at a tactical level, it seems that CE PSYOP would be the preferred option on proportionality calculations. As CE PSYOP cause no physical harm, it is always the preferred option. However, this may be wrong. CE PSYOP can indirectly cause physical harm, can cause psychological harm, directly and indirectly, and can cause social and political harm. Sustained cyber bullying increases the risk of suicide,[49] and it is no stretch to imagine that particular CE PSYOP campaigns could bring about similar physical and psychological harm. In addition, in line with the *jus ad bellum* proportionality points above, a CE PSYOP campaign can, and typically would, cause social and political unrest, instability, and long-term harms. We need to recognize here that such complications about proportionality are not unique to CE PSYOP; proportionality calculations are significantly complex[50] and require epistemic and moral judgements,[51] all of which are contested. The point here is not so much that these proportionality calculations should prevent the use of CE PSYOP. Rather, it is that these factors are again, complex, and need to be taken seriously and included in a thorough proportionality calculation

We also encounter a problem for discrimination. In particular, with CE PSYOP, given the exposure of civilians to ICTs, it is likely that many CE PSYOP campaigns will directly or inadvertently target civilians. On the traditional *jus in bello* discrimination criterion, civilians should not be targeted. Yet, with CE PSYOP, civilians are likely going to be the intended targets, or at least, be foreseen targets. As I have argued elsewhere, this draws out an issue where we might be forced to choose between the *jus in bello* proportionality and discrimination criteria.[52] The argument I develop there is not so much about how to choose between the two criteria, but more that when faced with a decision brought up by new technologies, our own preference is revealed to us. For some, they might believe that CE PSYOP are preferable to the use of kinetic force, ultimately favouring

---

49 Sameer Hinduja and Justin W. Patchin, "Connecting Adolescent Suicide to the Severity of Bullying and Cyberbullying," *Journal Of School Violence* 18, no. 3 (2019).
50 Thomas Hurka, "Proportionality in the Morality of War," *Philosophy And Public Affairs* 33, no. 1 (2005).
51 Adam Henschke, "Conceptualising Proportionality and Its Relation to Metadata," in *Intelligence and the Function of Government*, ed. Daniel Baldino and Rhys Crawley (Melbourne: Melbourne University Press, 2018).
52 "What Cyberweapons Tell Us About Our Just War," in *Ethics under Fire*, ed. Tom Frame (Sydney: UNSW Press, 2017).

proportionality over discrimination, despite the fact that this targets civilians and so violates the discrimination criterion. For others, they might believe that CE PSYOP ought not be used as civilians should not be knowingly or foreseeably targeted, despite kinetic operations likely to be less proportionate than CE PSYOP. Thus, on this view, discrimination counts more than proportionality. Again, my point here is not to offer guidance about which option is correct. Rather, it is to show that the ethics around CE PSYOP are murky, even when keeping things to a context of military ethics.

Can Defensive Cyber Enabled PSYOP Be Justified?

One of the further ethical challenges with CE PSYOP is that a decision to engage in aggressive CE PSYOP can lead to a comparable response from the adversary state and/or its people. In the original "Tallinn Manual," the authors explored the legality of countermeasures. Summarised, "Rule 9" of the Tallinn Manual states "A State injured by an internationally wrongful act may resort to proportionate countermeasures, including cyber countermeasures, against the responsible State." Therefore, the point here is that if State A's decision to use CE PSYOP is deemed to be a wrongful act, then State B may consider it has a right to engage in "proportionate countermeasures." Continuing from the discussion about discrimination, if State A's CE PSYOP actions have deliberately or foreseeably targeted State B's civilian population, then State B may consider that targeting State A's civilian population with CE PSYOP to be a legitimate countermeasure. The legality here is complex, and obviously a point where there is likely to be heavy disagreement. My position here is not to enter into discussions of international law. Rather it is to point out that CE PSYOP that target civilians can lead to one's own citizens being targeted under the principle of countermeasures. While proportionality is a consideration, the ethical challenge here extends beyond the risks of escalation. Instead, what we need to consider is that, if State A's own citizens are likely to be targets of CE PSYOP; can State A use CE PSYOP against its own people?

This raises significant political, legal, and ethical challenges for liberal democracies. The reason for this claim is that a state acting defensively may need or feel the need to use CE PSYOP on their own citizens to mitigate their adversaries' CE PSYOP. The basic idea is that effective CE PSYOP run by an adversary are not easily defeated by simple information campaigns.

Effective CE PSYOP will utilise affect and uses cognitive biases to influence the target populations' beliefs and actions. As the persistence of anti-vaxxer beliefs show, despite significant evidence and ongoing information to the contrary, the false beliefs are persistent and resilient. Central to this is that it is not enough to simply have a website or newsletter that presents facts that undermine an adversary's CE PSYOP, that information has to be crafted in ways that make it interesting and compelling. Such counter CE PSYOP will likely have to use PSYOP themselves. Further to this, the messengers are important in whether people will trust the message or not. This is one key lesson from community-oriented counter terrorism efforts: Those who have been terrorists or held extremist views and then recanted those views are likely to be better messengers than security agencies etc. "For counternarratives to be effective in contexts such as these, they must be posted by an entity that forum visitors believe to be genuine so as to limit the arousal of psychological reactance."[53] The point here is that the source of the counter CE PSYOP matters perhaps as much as the manner in which counter CE PSYOP are presented.

Once we recognize that liberal democracies might need to be engaged in CE PSYOP against their own citizens to counter an adversary's CE PSYOP, we encounter significant challenges. For instance, in the U.S. the U.S. Information and Educational Exchange Act of 1948, also known as the Smith-Mundt Act that authorizes government propaganda "not only promotes dissemination of truthful information, but it also restricts dissemination of government propaganda to Americans."[54] A 2020 clarification on domestic distribution of program material states that "[n]o funds authorized to be appropriated to the Department of State or the Broadcasting Board of Governors shall be used to influence public opinion in the United States."[55] Thus, there are legal constraints, requiring that U.S. government propaganda efforts must be truthful and not directed at U.S. citizens.

The underpinning political principle behind this is that liberal democracies do not engage in efforts to change the hearts and minds of their

---

[53] Allen W. Palmer and Edward L. Carter, "The Smith-Mundt Act's Ban on Domestic Propaganda: An Analysis of the Cold War Statute Limiting Access to Public Diplomacy," *Communication Law And Policy* 11, no. 1 (2006): 9.

[54] Kurt Braddock and John Horgan, "Towards a Guide for Constructing and Disseminating Counternarratives to Reduce Support for Terrorism," *Studies In Conflict & Terrorism* 39, no. 5 (2016): 392.

[55] Office Of The Law Revision Counsel Of The United States House Of Representatives, "22≈U.S.C § 1461 1a Clarification on Domestic Distribution of Program Material," (2013).

own people. In arguing for an amendment to the Smith-Mundt Act, senator Zerinksy stated "By law, the U.S. [Information Agency] cannot engage in domestic propaganda. This distinguishes us, as a free society, from the Soviet Union where domestic propaganda is a principal government activity."[56] One of the foundational principles of liberal democracies is that their citizens have a right to freely believe what they will, and efforts by the government agencies to change those beliefs are anathema to this principle. Thus, on this political principle, we ought not allow liberal democracies to engage in CE PSYOP against their own citizens. However, as we have just noted, perhaps a state needs to engage in CE PSYOP to counter an adversary's CE PSYOP. Added to this is that with the comprehensive integration of ICT into our lives, the citizens of liberal democracies are at greater risk of an adversary's CE PSYOP than they might have been to traditional PSYOP.

This is where the discussion of just war principles and white/black/grey taxonomy can become of practical use. Recall that there was complexity around the last resort criterion, and whether aggressive CE PSYOP could justifiably be used with a lower set of permissions than traditional warfare. Now, consider that a liberal democracy holds to the principle that it ought not use CE PSYOP against its own citizens. However, if it takes that principle seriously, it might have to reject calls to use CE against an adversary. The logic here is that, if State A reliably predicts that engaging in CE PSYOP against State B will motivate State B to use CE PSYOP against State A as a form of counter-measures, then State A will need to engage in CE PSYOP against its own citizens to counter State B's CE PSYOP. Thus, not only does State A's aggressive CE PSYOP predict a situation of escalation, but State A is now anticipating the need to use CE PSYOP against their own citizens. But on State A's own principles, they cannot do this; the restriction on directing CE PSYOP against their own people thus prevents them from engaging in aggressive CE PSYOP against State B.[57] While this is not an argument against states using aggressive CE PSYOP, it does raise an ethical challenge based in the state's own principles.

However, perhaps white CE PSYOP are permissible. One would certainly not reject the notion that a state not engage in any communicative efforts directed at its own citizens. I suggest here that truthful CE PSYOP

---

56 Government Publishing Office, "Congressional Record," Proceedings Of Congress And General Congressional Publications (Government Publishing Office, 1985).
57 I thank Ned Dobos for suggesting the construction of the problem in this way.

that are clear about the source might be ethically permissible to counteract an adversary's CE PSYOP. That said, this is a controversial position, and I am not wholly certain of its ethical permissibility. Any such domestically directed CE PSYOP would need to be considered very thoughtfully and would need to take into account not just the ethical, legal and political principles of the given state, but practical issues as well. For instance, any consideration of using CE PSYOP against a state's own citizens would need to think about what happens if it goes public. Decision makers would need to ask what impact would publicity of a counter CE PSYOP program have on the efficacy of this counter CE PSYOP program. Second, what impact would publicity of a counter CE PSYOP program have on public trust in the institution of government? Finally, what impact would publicity of a counter CE PSYOP program have on public trust in the elected government, specifically, the particular leaders and decision makers? The strength of white CE PSYOP is that these questions would need to be asked, and if conducted in an effective manner, the white CE PSYOP might not be detrimental either to the counter CE PSYOP effort, nor would it undermine trust in government or the political leadership.

Conclusion

This chapter has looked at the emerging challenge of CE PSYOP and suggested a range of ethical issues that need to be recognized and engaged with in order for CE PSYOP to be permitted. It opened by looking at the challenges faced by liberal democracies from adversaries remotely operated CE PSYOP. Having presented a description of PSYOP and CE PSYOP we were then able to lay out a taxonomy of CE PSYOP types. The just war tradition offers a useful set of principles to begin to analyse the ethical challenges around CE PSYOP. Perhaps unsurprisingly, when considering these principles, things become complex quite quickly. Though the just war criteria do not offer easy answers to whether CE PSYOP are permissible or not, the criteria are a useful set of tools by which to analyse and critically engage with questions of when CE PSYOP might be permissible. The chapter finished by looking at the legal, political, and ethical challenges of a liberal democratic state using CE PSYOP against its own citizens. Again, no easy answers are found, but if a state is to take a principle that it ought not use CE PSYOP against its own people, then it may in fact be constrained in engaging in aggressive CE PSYOP against an adversary.

Similarly, in recognising that liberal democratic states are indeed under attack from adversaries and may need to engage in CE PSYOP to protect their own people and democratic institutions, we saw that the idea of white CE PSYOP might be one way of resolving some of the legal, political, and ethical issues of using CE PSYOP against one's own citizens.

References

Allhoff, F., A. Henschke, and B. J. Strawser, eds. *Binary Bullets: The Ethics of Cyberwarfare*: (Oxford University Press, 2016).

Anspach, N. M. "The New Personal Influence: How Our Facebook Friends Influence the News We Read." *Political Communication* 34, no. 4 (2017/10/02 2017): 590-606.

Bellaby, R. *The Ethics of Intelligence*. Routledge, 2014. doi:doi:10.4324/9780203383575.

Braddock, K., and J. Horgan. "Towards a Guide for Constructing and Disseminating Counternarratives to Reduce Support for Terrorism." *Studies In Conflict & Terrorism* 39, no. 5 (2016): 381-404.

Brattberg, E., and T. Maurer. "Russian Election Interference: Europe's Counter to Fake News and Cyber Attacks." (Carnegie Endowmnent For International Peace, 2018).

Bricker, B., and J. Justice. "The Postmodern Medical Paradigm: A Case Study of Anti-Mmr Vaccine Arguments." *Western Journal Of Communication* 83, no. 2 (2019/03/15 2019): 172-89.

Caba, J. "Torture Methods with Sound: How Pure Noise Can Be Used to Break You Psychologically." *Medical Daily*, (20 January 2015).

Chairman Of The Joint Chiefs Of Staff. "Psychological Operations: Revision of Joint Publication 3-13.2." (Joint Chiefs Of Staff, 2010).

Clapper, J. R, and T. Brown. *Facts and Fears*. (Viking, 2019).

Coates, CAJ. *The Ethics of War*. (Manchester University Press, 1997).

Coleman, Stephen. *Military Ethics: An Introduction with Case Examples*. (New York: Oxford University Press, 2013).

Dobos, N. *Insurrection and Intervention: The Two Faces of Sovereignty*. (Cambridge University Press, 2011).

Dretske, F. I. *Knowledge and the Flow of Information*. (Oxford: B. Blackwell, 1981).

Evanina, W. "Statement by Ncsc Director William Evanina: Election Threat Update for the American Public." news release, (7 August, 2020), https://www.dni.gov/index.php/newsroom/press-releases/item/2139-statement-by-ncsc-director-william-evanina-election-threat-update-for-the-american-public.

Floridi, L. *The Philosophy of Information*. (Oxford: Oxford University Press, 2011).

———. "Semantic Conceptions of Information." In *The Stanford Encyclopedia of Philosophy*, edited by Edward N. Zalta, (2011).

Ford, S. B. *"Jus Ad Vim* and the Just Use of Lethal Force-Short-of-War." In *Routledge Handbook of Ethics and War: Just War in the 21st Century*, edited by Fritz Allhoff, Nicholas G. Evans and Adam Henschke. (Routledge, 2013).

Gibney, A. and J. Alberto Botero. "Agents of Chaos." United States: HBO, 2020.

Government Publishing Office. "Congressional Record." D3-D906: Government Publishing Office, (1985).

Griffin, A. "What Is Qanon? The Origins of Bizarre Conspiracy Theory Spreading Online." *Independent*, (24 August 2020).

Henschke, A. "Conceptualising Proportionality and Its Relation to Metadata." In *Intelligence and the Function of Government*, edited by Daniel Baldino and Rhys Crawley. (Melbourne: Melbourne University Press, 2018).

———. *Ethics in an Age of Surveillance: Virtual Identities and Personal Information*. (New York: Cambridge University Press, 2017).

———. "What Cyberweapons Tell Us About Our Just War." Chap. 14 In *Ethics under Fire*, edited by Tom Frame, 227-41. (Sydney: UNSW Press, 2017).

Hinduja, S. and J. W. Patchin. "Connecting Adolescent Suicide to the Severity of Bullying and Cyberbullying." *Journal Of School Violence* 18, no. 3 (2019/07/03 2019): 333-46.

Hurka, T. "Proportionality in the Morality of War." *Philosophy And Public Affairs* 33, no. 1 (2005): 34-66.

Intelligence And Security Committee Of Parliament. "Intelligence and Security Committee of Parliament: Russia." London: Intelligence And Security Committee Of Parliament, (2020).

Jowett, G. and V. O'Donnell. *Propaganda and Persuasion*. First ed. (Newbury Park: Sage Publications, 1986).

Kilcullen, D. *Blood Year: The Unvravelling of Western Counterterrorism*. (Oxford University Press, 2016).

Langner, R. "To Kill a Centrifuge: A Technical Analysis of What Stuxnet's Creators Wanted to Achieve." (37. Arlington: The Langner Group, 2013).

McMahan, J. *Killing in War*. (Oxford: Clarendon Press, 2009).

Miller, S. *The Moral Foundations of Social Institutions: A Philosophical Study*. (Cambridge: Cambridge University Press, 2010).

Miller, S. and J. Feltes. "Defining Terrorism." In *Counter-Terrorism: The Ethical Issues* edited by Seumas Miller, Adam Henschke and Jonas Feltes: (Edward Elgar, 2021).

Mueller, R. S. "United States of America V. Internet Research Agency." edited by United States Department Of Justice. District Of Columbia, (2018).

Office Of The Law Revision Counsel Of The United States House Of Representatives. "22 U.S.C § 1461 1a Clarification on Domestic Distribution of Program Material." (2013).

Omand, D. and M. Phythian. *Principled Spying: The Ethics of Secret Intelligence*. (Oxford University Press, 2018).

Orend, B. *The Ethics of War*. 2nd ed. (Vancouver: University Of Alberta, 2013).

Palmer, A. W., and E. L. Carter. "The Smith-Mundt Act's Ban on Domestic Propaganda: An Analysis of the Cold War Statute Limiting Access to Public Diplomacy." *Communication Law And Policy* 11, no. 1 (2006/01/01 2006): 1-34.

Perry, D. L. *Partly Cloudy: Ethics in War, Espionage, Covert Action, and Interrogation.* (Rowman & Littlefield, 2016).

Primoratz, I. "What Is Terrorism?". Chap. 2 In *Terrorism: The Philosophical Issues*, edited by Igor Primoratz, 15 - 27. (Basingstoke: Palgrave, 2004).

Reichberg, G., H. Syse, and E. Begby, eds. *The Ethics of War: Classic and Contemporary Readings*: (Blackwell, 2006).

Rid, T. *Cyber War Will Not Take Place.* Hurst & Company, 2013.

Robillard, M. "Counter-Terrorism Ethics and Psyop." In *Counter-Terrorism Ethics*, edited by Seumas Miller, Adam Henschke and Jonas Feltes: (Edward Elgar, Forthcoming).

Schmitt, M. N., ed. *Tallinn Manual on the International Law Applicable to Cyber Warfare*: (Cambridge, 2013).

Shearer, E. "Social Media Outpaces Print Newspapers in the U.S. As a News Source." news release, (10 December, 2018). https://www.pewresearch.org/fact-tank/2018/12/10/social-media-outpaces-print-newspapers-in-the-u-s-as-a-news-source/.

Smyczek, P. J. "Regulating the Battlefield of the Future: The Legal Limitations on the Conduct of Psychological Operations (Psyop) under Public International Law." *Air Force Law Review* 57 (2005): 209.

Thakur, R. "The Responsibility to Protect at 15." *International Affairs* 92, no. 2 (2016): 415-34.

Thorson, E. "Belief Echoes: The Persistent Effects of Corrected Misinformation." *Political Communication* 33, no. 3 (2016): 460-80.

United States of America V. Elena Alekseevna Khusyaynova, (2018).

Walzer, M. *Just War and Unjust Wars.* 4th ed. (New York: Basic Books, 2006).

PART THREE

9

THE MORALITY OF MACHINES

Cyber Guardian Angels

Andrew M. Tidmarsh

Introduction

The rapid pace of technological change and an increasingly pervasive information environment is transforming the character of warfare. Information, human perception, and psychology have always been a fundamental aspect of war and in many respects its key to success or failure; but the emergence of the cyber domain has opened a new frontier. This frontier is chaotic and changeable, at once moving at lightning speed, or painstakingly slowly; it is both accessible to all and its secrets exclusive to a few. Much of the discussion around the cyber domain reflects the conversations had about air power when it emerged and evolved through the twentieth-century: ubiquity, speed, technology, strategic effect. Yet the cyber domain is ubiquitous in a way that air is not. The presence of air as a medium does not remove the need to physically move a platform through it; in cyber, the existence of the medium provides its own speed-of-light platform. The exquisite, and expensive, technology required to dominate the air has limited its ownership to the powerful in any asymmetric contest. Not so with cyber, since cheap and ready access to computer systems and the internet offer an immediate path for almost anyone to make their presence felt.

The exertion of cyber power may be broadly grouped into two areas. On one hand, it enables. Whether through data science, means of production, sensors, weapons, or platforms: it enhances and informs our ability to operate in the physical world. On the other hand, it provides a pervasive medium to directly influence behaviour. We exist and breathe in the real world, but we perceive our own reality based on the information we receive, the way in which we receive it and manner in which the lens of our psychology focusses it for us. Consequently, the cyber domain offers the potential to manipulate behaviour more directly, and at a greater scale, than any other.

This chapter will discuss some of the risks emerging as a consequence of the changing character of conflict, particularly in relation to the fallibility of humans and the way in which some actors seek to capitalise on this by sowing discord with cyber-enabled information warfare. It will also highlight the threat of behavioural fratricide, where information "weaponised" against one's own population as the target audience for a political end, brings about unwanted and dangerous side effects. More optimistically, the potential for our behaviour to be supported by machines will be highlighted. Might it be possible that artificial intelligence (AI) could monitor the compound influence of behavioural risk factors in a way that may support our ethical decision making in the cyber, or any other, domain—a kind of cyber guardian angel?

Individual Personality

Fundamentally, humankind has not changed for tens of thousands of years. Our Stone Age ancestors possessed the same physical, emotional and intellectual abilities that we do.[1] The few thousand years over which humankind has developed the collective fictional structures upon which civilisation is based is a brief period in comparison to the tens of thousands of years humans spent as hunter-gatherers. From an evolutionary perspective, human psychology is the same now as it was in prehistory.[2] The development of the cyber domain has occurred in the blink of an eye and so the psychological lens through which we perceive the information provided to us through cyber space is not changing any time soon.

As humans, we are all unique. However, between us as a species, we also share common facets and personality traits that serve to make us astonishingly predictable, particularly when considering the prevailing behaviour of particular groups. Recent research conducted by Lindén et al. on Swedish army peacekeepers has suggested it is possible to predict moral

---
1 Yuval Noah Harari, *Sapiens: A Brief History of Humankind* (London: Penguin, 2015), 44.
2 Harari, *Sapiens*, 45.

transgressions on operations based on the presence of malevolent individual difference factors, specifically: the "Dark Triad" of Machiavellianism, narcissism and psychopathy; and socio-political attitudes relating to social-dominance orientation (SDO) and right-wing authoritarianism (RWA).[3] Importantly, participants in the research were not full-blown psychopaths, but represented the typical demographics of the Swedish, and western, armed forces. All of us possess varying combinations of these traits and in spite of our ability to flexibly think around the issues we face in the world, these and other factors influence how we interpret and respond to the information we receive.

The Dark Triad was introduced by Delroy Paulhus and Kevin Williams to describe the interrelated but distinct personalities of Machiavellianism, narcissism and psychopathy.[4] Individuals who score more highly for Dark Triad personality traits tend to experience less empathy, are more utilitarian in pursuing their goals and are more inclined to inter-group prejudice.[5] Higher scores in these personality traits has also been shown to be related to a higher endorsement of violence in war and an increased tolerance for the killing of civilians.[6]

Up to 2 percent of the general population would be diagnosed as either clinical psychopaths or narcissists, however the prevalence of subclinical psychopathy, subclinical narcissism and Machiavellianism can be expected to be much higher.[7] Studies conducted by MacManus and Messer suggest that the prevalence of these traits within the military may be twice that of

---

3 Magnus Lindén et al, "A latent core of dark traits explains individual differences in peacekeepers' unethical attitudes and conduct," *Military Psychology* 31, no. 6 (2019): 499-509.
4 Delroy L. Paulhus and Kevin M. Williams, "The Dark Triad of personality: Narcissism, Machiavellianism, and psychopathy," *Journal of Research in Personality* 36, issue 6 (December 2002): 556-563.
5 Gordon Hodson, Sarah M. Hogg and Cara C. MacInnis, "The role of "dark personalities" (narcissism, Machiavellianism, psychopathology), big five personality factors, and ideology in explaining prejudice," *Journal of Research in Personality* 43 (2009): 686–690, DOI:10.1016/j.jrp.2009.02.005. D. M. Bartels and D. A. Pizarro, "The mismeasure of morals: Antisocial personality traits predict utilitarian responses to moral dilemmas," *Cognition* 121 (2011): 154-161, DOI:10.1016/j.cognition.2011.05.010.
6 Victoria Blinkhorn, Minna Lyons and Louise Almond, "Drop the bad attitude! Narcissism predicts acceptance of violent behaviour," *Personality and Individual Differences* 98 (August 2016): 157-161, DOI:10.1016/j.paid.2016.04.025.
7 Jana Volkert, Thorsten-Christian Gablonski and Sven Rabung, "Prevalence of personality disorders in the general adult population in Western countries: systematic review and meta-analysis," *The British Journal of Psychiatry* 213, issue 6 (December 2018): 709-715, DOI: 10.1192/bjp.2018.202.

the general population.⁸ Whether we like it or not, our personality affects the way in which we interpret and respond to information. That said, gender differences aside, the distribution of these traits throughout the population make them relatively difficult to target for manipulation by information operations, short of casting a wide net and using barrage techniques. Sociopolitical attitudes are arguably easier to uncover and consequently easier to target.

Lindén's study concentrated on two socio-political attitudes that served to predict unethical attitudes and behaviour: SDO and RWA. SDO was defined by Jim Sidanius and Felicia Pratto as describing an individual's attitude towards non-egalitarian and hierarchical structured social orders.⁹ Individuals high in SDO have a preference for dominance hierarchies, including for systems where high-status groups forcefully oppress low-status groups.¹⁰ RWA comprises three related attitudes: authoritarian submission, authoritarian aggression and conventionalism.¹¹ Individuals high in RWA tend to organise their worldview in terms of in-groups and out-groups and perceive out-groups as threatening their traditional values.¹² They are generally self-righteous and manifest prejudice as a means to express aggression aroused by perceived threats, especially when out-groups are condemned by authority figures.¹³ RWA negatively predicts endorsement of human rights and strongly predicts support for human rights restrictions.¹⁴

SDO and RWA are closely related and both predict prejudice. Individuals high in SDO and RWA are more likely to adopt and maintain

---

8 Stephen C. Messer et al, "Projecting mental disorder prevalence from national surveys to populations-of-interest: An illustration using ECA data and the U.S. Army," *Social Psychiatry and Psychiatric Epidemiology* 39, issue 6 (June 2004): 419-426, DOI:10.1007/s00127-004-0757-1. Deirdre MacManus et al, "Impact of pre-enlistment antisocial behaviour on behavioural outcomes among UK military personnel," *Social Psychiatry and Psychiatric Epidemiology* 47 (2012): 1353-1358, DOI:10.1007/s00127-011-0443-z. Deirdre MacManus et al, "Violent offending by UK military personnel deployed to Iraq and Afghanistan: a data linkage cohort study," *The Lancet* 381 (April 2013): 907-917.
9 Lindén et al, "A latent core of dark traits," 501.
10 Arnold K. Ho et al, "The nature of social dominance orientation: Theorizing and measuring preferences for intergroup inequality using the new SDO₇ scale," *Journal of Personality and Social Psychology* 109, issue 6 (2015): 1003-1028.
11 Bob Altemeyer, "The other 'authoritarian' personality," *Advances in Experimental Social Psychology* 30 (1998): 47-92.
12 Bernard E. Whitley, "Right-Wing Authoritarianism, Social Dominance Orientation, and Prejudice," *Journal of Personality and Social Psychology* 77, no. 1 (1999): 126-134.
13 Whitley, "Right-Wing Authoritarianism."
14 J. Christopher Cohrs et al, "Determinants of Human Rights Attitudes and Behavior: A Comparison and Integration of Psychological Perspectives," *Political Psychology* 28, no. 4 (2007): 458.

factually inaccurate beliefs and misperceptions that support existing prejudices.[15] What does this mean for information operations in cyber space?

Exerting Influence Through Cyber Space

The concept of "us" and "them" is a powerful motivator that harnesses humans' natural tendency to support and defend their community against external threats. As previously discussed, certain personality types and socio-economic attitudes make some individuals more susceptible to latent prejudice that can emerge or be exacerbated by the stirring of these emotions.

Social Identity Theory describes how individuals typically define themselves in terms of their social identity; they attach value to their in-group because their sense of self is implicated within it.[16] Social identity is the basis of social influence. In-group members are perceived to be better qualified to provide information about self-relevant features of reality and we are motivated to coordinate our behaviour on these matters.[17] Individuals respond most strongly to perceived expert authority within their in-group; most powerfully exerted by those who best embody the identity of the group, the prototypical in-group member.[18] Thus, group members may be strongly influenced to act against an out-group, especially when they are motivated by service to the in-group's ideology and where the out-group is socially distant. In effect, emotional empathy for the out-group is suppressed. Ian Kershaw has argued that the dynamism of the Nazi regime came about not because its members were directly following orders, but rather because they were working creatively in ways they perceived the leadership would approve.[19] Its adherents identified more closely with the authority of the Nazi regime than with any alternate, more distant, locus of identification and were motivated by enthusiasm for the cause.[20]

---

15 Philip T. Dunwoody et al, "Authoritarianism, social dominance and misperceptions of war," *Peace and Conflict: Journal of Peace Psychology* 20, issue 3 (August 2014): 256-266.
16 Haslam, S. Alexander, and Stephen D. Reicher. "50 Years of 'Obedience to Authority': From Blind Conformity to Engaged Followership." *Annual Review of Law and Social Science* 13 (2017): 59-78.
17 Alexander and Reicher, "50 years of 'Obedience to Authority.'"
18 Michael A. Hogg, "A Social Identity Theory of Leadership," *Personality and Social Psychology Review* 5, no. 3 (2001): 184-200.
19 Ian Kershaw, "Working towards the Führer: reflections on the nature of the Hitler dictatorship," *Contemporary European History* 2 (1993): 103.
20 Claudia Koonz, *The Nazi Conscience* (New Haven, CT: Harvard University Press, 2003), passim.

The exploitation of social identity is a powerful tool that can be easily wielded through cyber space to manipulate behaviour and sow discord. Studies have indicated that over 150,000 Russian twitter accounts were active in the run-up to the UK's referendum on European Union (EU) membership, posing as British and posting in English, including urging the UK to "make June the 23rd our Independence Day," and generally applying nativist, often anti-Muslim rhetoric.[21] Narratives in social media can often take on a life of their own. A Symantec Threat Intelligence investigation into Russian social media involvement in the U.S. 2016 Presidential election found that Russia's Internet Research Agency (IRA) ran nearly 4000 twitter accounts, which amassed 6.4 million followers.[22] The most popular account, masquerading as a Republican party group in Tennessee, gained 150,000 followers and garnered six million retweets, only 1850 of which were from other IRA accounts. The implication being that the fake account's narrative was strongly amplified within its target audience.

Collaborative filtering algorithms are used by recommender systems to present users with content they are expected to approve of, based on their historical preferences and the assumption that they will continue to agree with those they have agreed with in the past.[23] On one hand, this results in excellent movie recommendations, but on the other, it can create "filter bubbles," where individuals find themselves in an echo chamber of political discourse that serves to reinforce social identity groups and exacerbate prejudice.[24] Although implemented for good reasons, taking advantage of the machine learning algorithms that classify us into groups based on our preferences is very easy to do, especially when equipped with some knowledge of human psychology. Bots can be easily programmed that work within social media systems, riding their algorithms and human

---

[21] United States Senate, "Putin's Asymmetric Assault on Democracy in Russia and Europe: Implications for U.S. National Security," Committee on Foreign Relations, 10 January 2018, https://www.foreign.senate.gov/imo/media/doc/FinalRR.pdf. David D. Kirkpatrick, "Signs of Russian Meddling in Brexit Referendum," The New York Times, 15 November 2017, https://www.nytimes.com/2017/11/15/world/europe/russia-brexit-twitter-facebook.html.
[22] Gillian Cleary, "Twitterbots: Anatomy of a Propaganda Campaign," Symantec Threat Intelligence, 5 June 2019, https://symantec-enterprise-blogs.security.com/blogs/threat-intelligence/twitterbots-propaganda-disinformation.
[23] Shuyu Luo, "Introduction to Recommender Systems," Towards Data Science, 10 December 2018, https://towardsdatascience.com/intro-to-recommender-system-collaborative-filtering-64a238194a26.
[24] Yong Min et al, "Endogenetic structure of filter bubble in social networks," *Royal Society Open Science* 6, no. 11 (November 2019): 190868.

behaviours online to amplify their message.²⁵ This is a good example of the ease with which the cyber domain may be exploited by those with technical knowhow against those with ready access, but who are ill-prepared.

Behavioural Fratricide

One problem with attempting to bring about behavioural change through information operations in the cyber domain is that the same accessibility and ubiquity that makes the medium effective, also makes it an imprecise and unpredictable tool. Whilst artificially intelligent systems can be used to customise information feeds to target the individual, it is group behaviour that they aim to affect. In doing so, they can often bring about unexpected or unintended results. Russia's interference in the U.S. elections was predominantly in support of Donald Trump, but not exclusively. What was consistent, was the intention to sow division. Targeting the Republican base was more likely to create a larger effect as a result of the susceptible personality traits associated with right-wing views and conservatism, but it could be argued that the actual election of Trump was a by-product. The direct effect of information operations is extremely hard to quantify, but Russia's leadership is likely to feel pleased that it has contributed to increasingly stark and challenging divisions that are growing in the U.S. today.²⁶

A classic appeal for action to support the in-group against an out-group is that of nationalism. Inherently adversarial, its danger has been demonstrated by the identity politics surrounding the UK's departure from the EU. The campaign to leave the EU was explicitly nationalist and overtly anti-immigration. Often this manifested in Islamophobia, even though at face value this should have been entirely unrelated. A significant body of evidence has suggested that prejudice played a large role in the outcome of the referendum.²⁷ In particular, research has shown that collective

---

25 Fabian Bosler, "Twitter — Or where my bot talks to your bot," Towards Data Science, 12 October 2019, https://towardsdatascience.com/twitter-3478a68d5875.
26 Michael Dimock and Richard Wike, "America is exceptional in the nature of its political divide," Pew Research Center, 13 November 2020, https://www.pewresearch.org/fact-tank/2020/11/13/america-is-exceptional-in-the-nature-of-its-political-divide/.
27 Paul B. Hutchings and Katie E. Sullivan, "Prejudice and the Brexit vote: a tangled web," *Palgrave Communications* 5, Article 5 (2019), https://doi.org/10.1057/s41599-018-0214-5. Tina G. Patel and Laura Connelly, "'Post-race' racisms in the narratives of 'Brexit' voters," *The Sociological Review* 67, issue 5 (September 2019): 968-984.

narcissism, RWA and SDO were independently related to the perceived threat of immigrants and in turn strongly predicted support for leaving the EU, over and above other notable variables such as education and age.[28] The Leave campaign and supporting media exploited prejudice and fear of the out-group.[29] The campaign's success demonstrated the combined power of individual disposition, social identity and systemic encouragement, often through cyber space.

There are dangers to wielding identity politics in this manner. Studies in 2018 and 2019 showed that half of British people felt Islam was fundamentally incompatible with British values and they would not accept Muslims as family members.[30] Recent research has shown that hate-speech by political figures boosts domestic terrorism.[31] Racial violence and hate-speech in the UK increased markedly following the EU referendum, which research has attributed to a racist climate encouraged by the nativist narrative of the referendum debate and previous divisive government policies.[32]

A similar effect can generally be seen whenever an out-group is highlighted or linked with threat. During the COVID-19 pandemic, "combating" COVID-19 was typically framed in the language of war.[33] Partly as a result, nearly every democratic leader saw an initial increase in approval rating, despite differing strategies and levels of success in handling the virus.[34] However, one cannot "fight" a virus in the physical

---

28 Agnieszka Golec de Zavala, Rita Guerra and Cláudia Simão, "The Relationship between the Brexit Vote and Individual Predictors of Prejudice: Collective Narcissism, Right Wing Authoritarianism, Social Dominance Orientation," *Frontiers in Psychology* 27 (November 2017): 2023, DOI=10.3389/fpsyg.2017.02023.

29 Martin Moore and Gordon Ramsay, "UK media coverage of the 2016 EU Referendum campaign," King's College London, Centre for the Study of Media, Communication and Power, May 2017, https://www.kcl.ac.uk/policy-institute/assets/cmcp/uk-media-coverage-of-the-2016-eu-referendum-campaign.pdf.

30 Pew Research Center, "Being Christian in Western Europe," Religion and Public Life, 29 May 2018, https://www.pewforum.org/2018/05/29/being-christian-in-western-europe/. Pew Research Center, "In the U.S. and Western Europe, people say they accept Muslims, but opinions are divided on Islam," Fact Tank News in Numbers, 8 October 2019, https://www.pewresearch.org/fact-tank/2019/10/08/in-the-u-s-and-western-europe-people-say-they-accept-muslims-but-opinions-are-divided-on-islam/.

31 James A. Piazza, "Politician hate speech and domestic terrorism," *International Interactions* (March 2020), DOI: 10.1080/03050629.2020.1739033.

32 Jon Burnett, "Racial Violence and the Brexit State," *Race and Class* 58, issue 4 (2017): 85-97.

33 Lawrence Freedman, "Coronavirus and the Language of War," New Statesman, 11 April 2020, https://www.newstatesman.com/science-tech/2020/04/coronavirus-and-language-war.

34 Kiran Stacey and Jim Pickard, "Coronavirus pandemic boosts popularity of Trump and Johnson," Financial Times, last updated 31 March 2020, https://www.ft.com/content/c7f5a8bc-eb0e-45e5-a080-bbfd6d317def.

sense and stigmatizing language surrounding COVID-19 has led to a surge in hate crime against ethnic-Asian groups. On 11 February 2020, the World Health Organisation advised that political leaders and media institutions should not "attach locations or ethnicity to the disease, this is not a 'Wuhan Virus,' 'Chinese Virus' or 'Asian Virus.' The official name for the disease was deliberately chosen to avoid stigmatisation."[35] Nevertheless, President Trump and other prominent individuals in his administration did commonly refer to the virus as the "China Virus." Research published at the end of 2020 has shown that this was associated with an increase in implicit bias against Asian Americans.[36] In the UK, between January and March 2020, hate-crime against ethnic-Asian people increased by nearly 300 percent.[37]

Right or Wrong?

Ethics is a quintessentially human field with a number of interconnected and contesting schools, spanning theology, philosophy, psychology and neuroscience. Our perception of right and wrong is at least to some degree relative, varying with culture and time, which demonstrates a socio-cultural learned basis for morality. However, there are also commonalities that persist through time and across groups, suggesting a deeper foundation of fundamental principles. At the same time, these principles clearly favour our own species over others, implying an evolutionary basis for morality.

Joshua Greene argues for a dual-process model of morality, built upon the two-system model proposed from the work by Daniel Kahneman and Amos Tversky.[38] In this model, automatic emotional processes dominate deontological decisions, whereas controlled cognitive processes largely influence utilitarian decisions. Amongst the range of Greene's research is an exploration of myriad variations of the famous *trolley problem*, including

---

[35] World Health Organisation, "Social Stigma associated with COVID-19," 24 February 2020, https://www.who.int/docs/default-source/coronaviruse/covid19-stigma-guide.pdf.
[36] Sean Darling-Hammon et al, "After "The China Virus" Went Viral: Racially Charged Coronavirus Coverage and Trends in Bias Against Asian Americans," *Health Education and Behaviour* 47, issue 6 (September 2020): 870-879.
[37] David Mercer, "Coronavirus: Hate crimes against Chinese people soar in UK during COVID-19 crisis," Sky News, 5 May 2020, https://news.sky.com/story/coronavirus-hate-crimes-against-chinese-people-soar-in-uk-during-covid-19-crisis-11979388.
[38] Daniel Kahneman, *Thinking, Fast and Slow* (London: Penguin, 2012). Joshua Greene, *Moral Tribes: Emotion, Reason and the Gap between Us and Them* (London: Atlantic Books, 2015).

the use of functional magnetic resonance imaging (fMRI) of participants' brains to interpret the differing ways moral decisions are being taken.

In the basic version of the problem, a person is faced with a dilemma: standing on a bridge, they see a rogue trolley hurtling towards a group of people, who will certainly be killed if they do not act. Thinking quickly, they realise that the only way to save the lives of the group is to push a stranger, who happens to be stood there with them, off the bridge. Doing so will kill the stranger, but also stop the trolley and save more lives. The person could jump, but an astute assessment of relative weight and some fast maths clearly indicates that the stranger, who is wearing a large rucksack, will be substantial enough to stop the trolley, whereas they would not. When asked, most people would say that this is wrong and, in sparing the stranger, it is therefore "right" to allow to trolley to continue its path and watch it kill the group.

In a similar problem, our person in question finds themselves at a railway switch box just in time to see an approaching runaway train. They know that a work group is on the line, just up the track, and if the train is allowed to continue it will certainly kill them. They also know that if they switch the tracks the train will divert away from the group, saving them, but will inevitably kill an individual who happens to be on the branch fork. Most people will say that this is the right thing to do. The outcomes and consequences of these two decisions are identical, so why does the typical automatic human response differ? Why is it more wrong to kill someone in person than to do so with a switch?[39]

It is unlikely that participants in this thought experiment undertake a comprehensive moral and philosophical analysis in the moments before they make their choice, their decision is simply based on what they feel, or do not feel. Greene's research suggests that the answer to morality lies in the brain, in particular the ventromedial prefrontal cortex (vmPFC), which plays a key role in the regulation of emotion.[40] When we are triggered to respond emotionally, our perception of what is right and wrong changes. These triggers can be subtle and varied and whilst the trolley problem in isolation may seem like an unrealistic curiosity, it has led to a broad array of research that demonstrates the inconsistency of our moral decision making. Our feelings over what is right and wrong are impacted by physical distance, emotional stimuli, tiredness, time pressure, whether

---

39 Greene, *Moral Tribes*, 116.
40 Greene, *Moral Tribes*, 116-128.

or not individuals are identifiable, the number of people in question. Perversely, the greater the number of people in need, the less we care; especially if they happen to be far away and we don't know their names. The point here, is that human morality is inconsistent. We are hard wired to respond emotionally in favour of local individuals that we know, because it engenders societal cooperation, which in turn helps our tribe to thrive against external competition.[41] Indeed, we know from our understanding of the Dark Triad that if we lack personal empathy, à la psychopaths and narcissists, then inter-personal relationships can go very badly wrong. We just don't extend this empathy far beyond our local sphere.

The argument is that the automatic processes that drive our fundamental feelings of right and wrong have evolved as a means to encourage in-group cooperation, but they often fail us when we interact with out-groups with whom we are predisposed to compete. When we engage our ability for flexible thinking, we can shift our position, but strangers with different viewpoints in the farthest reaches of cyberspace are least likely to stir our empathy or concern.[42]

The ubiquity of the cyber domain, coupled with ever increasing computing power, the availability of vast amounts of data and elegantly crafted algorithms to divine insight, has clearly opened up a new and significant threat vector. There is no going back from this: the genie is out of the lamp. However, these same developments also provide a means for defence. One approach to this may be an exploration of what it means to be moral and the potential of machines to understand and appreciate it.

Recognising the systematic nature of Greene's model, it may be possible to leverage the increasing capability of AI to teach, reflect or artificially generate morality. If our sense of morality is learned, either through the evolutionary biology of our brains, or by the culture in which live, then it might be possible to train an AI to respond to situations morally as we would wish it to. Moreover, an AI might be *more* morally consistent than we are. And, if we felt it needed to improve, then it's architecture could be redesigned and retrained in a way that the physical structure of our brains cannot be. Doing so could create a guardian capability that may warn of increasingly malign influence, of impending immoral actions; or it could help us check our own information campaigns to prevent unwanted behavioural fratricide.

---

41 Greene, *Moral Tribes*, 261-262.
42 Joshua Greene, "The Cognitive Neuroscience of Moral Judgement and Decision Making," in *The Moral Brain: A Multidisciplinary Perspective*, ed. Jean Decety and Thalia Wheatley (London: MIT Press, 2015), 197–220.

Moral Machines

An artificial neural network models the processes involved when the brain is intelligently controlling thought.[43] A biological neuron receives input signals through dendrites into a cell body. If the accumulation of input signals exceeds a particular threshold, an output signal is generated through an axon and axon terminals to other neurons. An artificial neuron works in a similar way, receiving a number of inputs and producing an output based on the weight given to each input and the neuron's overall bias.[44] The weights can be thought of as representing the strength of synapses between biological neurons and convey the significance of each input. The bias relates to a neuron's threshold, making it more or less sensitive to firing. Each neuron can be thought of as making a simple decision. It follows that more neurons are able to make more complex decisions. An artificial neural network is comprised of multiple neurons in a number of layers. The neurons in hidden layers enable progressively more complex and abstract decisions to be made. Different layers can incorporate varying types of connections, activation functions or recurrence, which can be applied to introduce short-term memory in support of longer-term understanding. Training an artificial neural network involves strengthening or weakening potentially millions of weights and biases to minimise its loss, or prediction error: a form of artificial neuroplasticity. Modern computing power has enabled the use of deep neural networks to provide extraordinary insights. As that power increases, the scope of what is possible will too.

By way of example, convolutional neural networks (CNN) have proven extremely effective at image recognition tasks, in some cases surpassing human performance.[45] In simple terms, convolution involves passing filters of various sizes over the neurons of a network that represent the pixels of an image for each of its colour channels. This has the effect of emphasising features, for example edges, which may be amplified further by pooling and can be repeated to varying depths. When trained with large datasets, the CNN will come to learn distinctive features from different

---

43 Richard E. Neapolitan and Xia Jiang. *Artificial Intelligence: With an Introduction to Machine Learning*, Second Edition (Boca Raton, FL: CRC Press, 2018), 389.
44 Michael A. Nielsen, "Neural Networks and Deep Learning," Determination Press, last updated 26 December 2019, Chapter 1, http://neuralnetworksanddeeplearning.com/.
45 Kaiming He et al, "Delving Deep into Rectifiers: Surpassing Human-Level Performance on ImageNet Classification", Microsoft Research, *International Conference on Computer Vision (ICCV)* 2015: 1026-1034.

perspectives; for example of a cat or a car ... or a soldier or a civilian; or a perhaps a person who appeared to be a civilian but is now carrying a weapon ... Introducing recurrence into these networks enables them to remember and learn how features change with time and what that means, for example a cat jumping, a car turning ... or perhaps a soldier raising their weapon ...

An AI may look at a scene and be able to classify multiple objects, using multiple classes and even multiple labels, where an object's properties suggest membership of more than one class: for example, the civilian-clothed insurgent.[46] The classification of a scene does not need to be limited to physical objects, but could include non-physical concepts and context: movie genres; expressed emotions; the presence of threat or "right" and "wrong."

A Picture Paints a Thousand Words ... but a Thousand Words Can Paint Any Picture

AI image recognition has enormous potential, but as humans we know that a picture can easily be taken out of context, which has clear implications for moral decision making. We supplement our own image of the world with language, usually either aurally or in written form, whose exposition provides context and detail for us to weigh our perspective. In military terms, tactical communication is strict, concise and formatted; born of the necessity for unambiguous, clear communication in the moments where life and death hang in the balance:

"Group, bullseye, 0-9-0, 36, 14 thousand, hostile."

To those trained in the language, this short sentence expressed a picture in an air battle, describing: the number of airborne radar contacts seen, the lateral- and altitude-bounded distance between them; their position in three-dimensional space; the range of their speed; their aspect in relation to friendly aircraft; their combat identification; permissions with respect to rules of engagement; and the expected actions to be taken. The information contained in these formats is explicitly implicit and relatively straight forward for machines to learn.

---

46 Pulkit Sharma, "Build Your First Multi-Label Image Classification Model in Python," 15 April 2019, Analytics Vidhya, https://www.analyticsvidhya.com/blog/2019/04/build-first-multi-label-image-classification-model-python/.

Generally, however, interpretation of language is more subjective, it is neither a transparent nor a static medium. Different languages possess different emphasis and tone, which can alter interpretation, understanding and opinion. Some languages, including English, lack vocabulary to express the nuance of others. Language is always changing, in some ways becoming more simplistic, in others more complex. It varies across generations. The way in which I describe the civilian-clothed insurgent, terrorist, freedom fighter, rebel, resistance hero, criminal, martyr will affect how you feel about them, based on your own culture and background. The fact that I have switched to a first-person style will have, unbidden, generated an emotional connection that may influence how you react to the points I am trying to convey. It could even retrospectively affect how favourable your memory of this chapter is (although my pulling back the curtain rather ruins the spell…). It has even been argued by Beatrice Heuser that differing semantic interpretations of Carl von Clausewitz's *On War* can be identified echoing through the history of diverse theories on warfare, including in Western doctrine.[47]

Natural language processing is a branch of AI that seeks to divine meaning from language. Most of us should already be familiar with it: email providers use these techniques to classify whether or not an email is spam. If you have ever strayed into the dark reaches of your junk folder, you will see they are actually very effective.

To some extent, the capability to discern right from wrong in language already exists. In recent years, automatic detection systems have become more effective at classifying hate-speech. A number of machine learning techniques have proven capable of discerning whether online communication could be considered offensive, racist, sexist, hateful or aggressive.[48] This is not without challenges: the evolution of language and opinion, as well as attempts to mask hate-speech, mean that efforts to detect it must be constantly adapted.

"the merciless Indian Savages, whose known Rule of Warfare, is an undistinguished Destruction, of all Ages, Sexes and Conditions"

Is this hate speech? It is actually an extract from the U.S. Declaration of Independence. A historian citing the text would not necessarily endorse those views, but still have a legitimate reason to quote it.[49]

---

47 Beatrice Heuser, *Reading Clausewitz* (London: Pimlico, 2002), 114-116.
48 Sean MacAvaney et al, "Hate speech detection: Challenges and solutions," *PLoS ONE* 14, issue 8 (August 2019): e0221152, DOI: 10.1371/journal.pone.0221152.
49 MacAvaney, "Hate speech detection."

Nevertheless, improving capability will naturally lead to more sensitive and nuanced classification of language that may one day be more adept at detecting the potential impact of words than many humans. In the simplest sense, a machine's capacity for vocabulary already far outstrips that of a human being: the average native English speaker will be aware of somewhere between twenty-seven thousand and fifty-two thousand *lemmas*, or uninflected words—a drop in the memory ocean for a machine.[50]

Using other AI techniques, communities and filter bubbles may be mapped on social media, providing insights into demographics, key influencers and resonating themes.[51] Our understanding of the personalities and attitudes that identify certain groups as susceptible to influence and exposes them to targeting, could also be used to defend them.

Whose Morality is it, Anyway?

Were a moral machine to be realised, it is possible that these systems could be employed wherever there is a machine in the loop of human decision making, whether at the strategic level of information operations, or at the tactical level of physical warfighting.

Typically, the ethical debate over the development of autonomous military systems has focussed on the risk. The concern is often to have "meaningful human control" or a "person in the loop," an insurance policy for humanity, because machines are devoid of conscience.[52] There are certainly strong legal and ethical arguments for ensuring that human emotion, compassion and conscience have a place in decisions over life and death.[53] Yet, at the same time, our morality is changeable, not only between disparate groups, but also within ourselves. We are fallible. Therefore, it seems strange to emphasise the ethical pre-eminence of human decision-making over emotionless machines, without also considering the converse:

---

50 Marc Brysbaert et al, "How Many Words Do We Know? Practical Estimates of Vocabulary Size Dependent on Word Definition, the Degree of Language Input and the Participant's Age," *Frontiers in Psychology* 7:1116. doi: 10.3389/fpsyg.2016.01116.
51 Kyle Findlay and Ockert Janse van Rensburg, "Using interaction networks to map communities on Twitter," *International Journal of Market Research* 60, issue 2 (March 2018): 169-189. Yong Min et al, "Endogenetic structure of filter bubble in social networks."
52 Mary Ellen O'Connell, "Banning Autonomous Killing: The Legal and Ethical Requirement that Humans Make Near-Time Lethal Decisions," in *The American Way of Bombing: Changing Ethical and Legal Norms from Flying Fortresses to Drones*, ed. Matthew Evangelista and Henry Shue (Ithaca: Cornell University Press, 2014), *passim*.
53 O'Connell, "Banning Autonomous Killing."

humans can err where machines might not. Furthermore, by our own standards, perhaps one day, moral machines could be more upstanding and incorruptible than we are.

This does beg the question: to whose standards and for whose morality should we train our artificial guardian angels? The Just War Tradition provides a well-recognised framework, at least for the western way of war. However, it is also recognised that military personnel can face ethically insoluble dilemmas, where there is simply no "good" option to take. Consequently, agents engaged in armed conflict require something more than rule-based ethical reasoning.[54]

Militaries should be there to protect the societies that they serve. So perhaps the moral standard of the average member of that society should be reflected, based on a dataset of representative moral judgements made by such a person. At the same time, we also expect a higher standard from our serving personnel, entrusted as they are with the responsibility to uphold and protect the security of our, at least declared, values. Furthermore, as part of their professional education, military personnel undergo ethics training, something perhaps few civilians will do. This has been shown to positively influence how soldiers interact with non-combatants and reduce unethical battlefield conduct.[55] So, the standard of the average professional military person. This seems reasonable. After all, it is what we expect of our human guardians.

Once again, we come face to face with the *trolley problem*. When it comes to machines, we often demand a significantly higher standard than we can achieve ourselves. In the U.S. in 2019, there were approximately thirty-six thousand fatal road traffic collisions.[56] Drunkenness, distraction and tiredness are factors in over half of all fatal accidents, with other forms of human error accounting for the vast majority.[57] The qualitative and early quantitative evidence suggests that driverless vehicles are safer than human-driven vehicles, yet 43 percent of people who are familiar with

---

54 Marcus Schulzke, "Ethically insoluble dilemmas in war," *Journal of Military Ethics* 12, no. 2: 95-110.
55 Christopher H. Warner et al, "Effectiveness of Battlefield-Ethics Training During Combat Deployment: A Programme Assessment," *The Lancet* 378, issue 9794 (September 2011): 915.
56 Alex Kopestinsky, "How Many People Die In Car Accidents?" Policy Advice, 27 February 2021, https://policyadvice.net/insurance/insights/car-accidents/.
57 Nidhi Kalra, Susan M. Paddock, "How Many Miles of Driving Would It Take to Demonstrate Autonomous Vehicle Reliability?" Driving to Safety, RAND Corporation, 2016, https://www.rand.org/pubs/research_reports/RR1478.html.

them do not feel safe in them.[58] There is also an ongoing ethical debate about the implications of a self-driving machine taking the life of a human being. In trolley terms, we emphasise the need to save the individual from a fatal accident caused by a machine, over the cost of potentially thousands of lives caused by the driving of human beings.

Automation is already prevalent in modern military capability. As it continues to develop and the scope of autonomous machine agency expands, it is quite likely that an analogous debate will (continue to) emerge; even if the evidence suggests that machines could act with greater moral precision than their human counterparts. Already, with a few exceptions, the debate mostly focuses on the tactical level of warfare, where the trigger is pulled, conjuring the most visceral images of life and death. This is the direct, proximate trigger for our emotional response, the one that causes us to care more about the local individual we know and less about the many who happen to be further away. Yet, we have already unleashed AI to make autonomous targeting decisions in the cyber domain for information operations. According to our own doctrine, this represents the strategic level of war, with more profound implications for our security and impacting many more thousands, or even millions of people, just less directly.[59]

Ironically, when it comes to the creation of moral machines, the greatest challenge may not lie in handling machine morality, but our own.

References

Altemeyer, B. "The other 'authoritarian' personality." *Advances in Experimental Social Psychology* 30 (1998): 47-92.

Bartels D. M., and D. A. Pizarro. "The mismeasure of morals: Antisocial personality traits predict utilitarian responses to moral dilemmas." *Cognition* 121 (2011): 154-161. DOI:10.1016/j.cognition.2011.05.010.

Blinkhorn, V., M. Lyons and L. Almond. "Drop the bad attitude! Narcissism predicts acceptance of violent behaviour." *Personality and Individual Differences* 98 (August 2016): 157-161. DOI:10.1016/j.paid.2016.04.025.

Bosler, F. "Twitter—Or where my bot talks to your bot." *Towards Data Science*, 12 October 2019. https://towardsdatascience.com/twitter-3478a68d5875.

---

58 Alex Kopestinsky, "25 Astonishing Self-Driving Car Statistics for 2021" Policy Advice, 29 April 2021, https://policyadvice.net/insurance/insights/self-driving-car-statistics/.
59 UK Ministry of Defence, "The Integrated Operating Concept 2025," 30 September 2020, https://www.gov.uk/government/publications/the-integrated-operating-concept-2025.

Brysbaert, M., M. Stevens, P. Mandera and E, Keuleers. "How Many Words Do We Know? Practical Estimates of Vocabulary Size Dependent on Word Definition, the Degree of Language Input and the Participant's Age." *Frontiers in Psychology* 7:1116. doi: 10.3389/fpsyg.2016.01116.

Burnett, J. "Racial Violence and the Brexit State." *Race and Class* 58, issue 4 (2017): 85-97.

Cleary, G. "Twitterbots: Anatomy of a Propaganda Campaign." Symantec Threat Intelligence, 5 June 2019. https://symantec-enterprise-blogs.security.com/blogs/threat-intelligence/twitterbots-propaganda-disinformation.

Cohrs, J. C., J. Maes, B. Moschner and S. Kielmann. "Determinants of Human Rights Attitudes and Behavior: A Comparison and Integration of Psychological Perspectives." *Political Psychology* 28, no. 4 (2007): 441-470.

de Zavala, A.G., R. Guerra and C. Simão. "The Relationship between the Brexit Vote and Individual Predictors of Prejudice: Collective Narcissism, Right Wing Authoritarianism, Social Dominance Orientation." *Frontiers in Psychology* 27 (November 2017): 2023. DOI=10.3389/fpsyg.2017.02023.

Darling-Hammond, S., E. K. Michaels, A. M. Allen, D. H. Chae, M. D. Thomas, T. T. Nguyen, M. M. Mujahid and R. C. Johnson. "After 'The China Virus' Went Viral: Racially Charged Coronavirus Coverage and Trends in Bias Against Asian Americans." *Health Education & Behavior* 47, issue 6 (September 2020): 870-879. DOI: 10.1177/1090198120957949.

Dimock, M. and R. Wike. "America is exceptional in the nature of its political divide." *Pew Research Center*, 13 November 2020. https://www.pewresearch.org/fact-tank/2020/11/13/america-is-exceptional-in-the-nature-of-its-political-divide/.

Dunwoody, P. T., D. L. Plane, S. A. Trescher and D. Rice. "Authoritarianism, social dominance and misperceptions of war." *Peace and Conflict: Journal of Peace Psychology* 20, issue 3 (August 2014): 256-266.

Findlay, K. and O. Janse van Rensburg. "Using interaction networks to map communities on Twitter." *International Journal of Market Research* 60, issue 2 (March 2018): 169-189.

Freedman, L. "Coronavirus and the Language of War." New Statesman, 11 April 2020. https://www.newstatesman.com/science-tech/2020/04/coronavirus-and-language-war.

GOV.UK. "The Integrated Operating Concept 2025," 30 September 2020. https://www.gov.uk/government/publications/the-integrated-operating-concept-2025.

Greene, J. *Moral Tribes: Emotion, Reason and the Gap between Us and Them.* London: Atlantic Books, 2015.

Greene, J. "The Cognitive Neuroscience of Moral Judgement and Decision Making." In *The Moral Brain: A Multidisciplinary Perspective*, edited by Jean Decety and Thalia Wheatley, 197–220. London: MIT Press, 2015.

Harari, Y.N. *Sapiens: A Brief History of Humankind.* London: Vintage, 2015.

Haslam, S. A. and S. D. Reicher. "50 Years of 'Obedience to Authority': From Blind Conformity to Engaged Followership." *Annual Review of Law and Social Science* 13 (2017): 59-78.

He, K., X. Zhang, S. Ren and J. Sun. "Delving Deep into Rectifiers: Surpassing Human-Level Performance on ImageNet Classification." Microsoft Research. International Conference on Computer Vision (ICCV), 2015: 1026-1034.

Heuser, B. *Reading Clausewitz*. London: Pimlico, 2002.

Ho, A. K., J. Sidanius, N. Kteily, J. Sheehy-Skeffington, F. Pratto, K. E. Henkel, R. Foels, A. L. Stewart. "The nature of social dominance orientation: Theorizing and measuring preferences for intergroup inequality using the new SDO$_7$ scale." *Journal of Personality and Social Psychology* 109, issue 6 (2015): 1003-1028.

Hodson, G., S. M. Hogg and C. C. MacInnis. "The role of "dark personalities" (narcissism, Machiavellianism, psychopathology), big five personality factors, and ideology in explaining prejudice." *Journal of Research in Personality* 43 (2009): 686–690. DOI:10.1016/j.jrp.2009.02.005.

Hogg, M. A. "A Social Identity Theory of Leadership." *Personality and Social Psychology Review* 5, no. 3 (2001): 184-200.

Hutchings, P. B. and K. E. Sullivan. "Prejudice and the Brexit vote: a tangled web." *Palgrave Communications* 5, Article 5 (2019). https://doi.org/10.1057/s41599-018-0214-5.

Kahneman, D. *Thinking, Fast and Slow*. London: Penguin, 2012.

Kalra, N. and S. M. Paddock, "How Many Miles of Driving Would It Take to Demonstrate Autonomous Vehicle Reliability?" Driving to Safety, RAND Corporation, 2016, https://www.rand.org/pubs/research_reports/RR1478.html.

Kershaw, I. "Working towards the Führer: reflections on the nature of the Hitler dictatorship." *Contemporary European History* 2 (1993): 103-8.

Kirkpatrick, D. D. "Signs of Russian Meddling in Brexit Referendum." *The New York Times*, 15 November 2017. https://www.nytimes.com/2017/11/15/world/europe/russia-brexit-twitter-facebook.html.

Koonz, C. *The Nazi Conscience*. New Haven, CT: Harvard University Press, 2003.

Kopestinsky, A. "How Many People Die In Car Accidents?" *Policy Advice*, 27 February 2021. https://policyadvice.net/insurance/insights/car-accidents/.

Kopestinsky, A. "25 Astonishing Self-Driving Car Statistics for 2021." *Policy Advice*, 29 April 2021. https://policyadvice.net/insurance/insights/self-driving-car-statistics/.

Lindén, M., F. Björklund, M. Bäckström, D. Messervey and D. Whetham. "A latent core of dark traits explains individual differences in peacekeepers' unethical attitudes and conduct." *Military Psychology* 31, no. 6 (2019): 499-509.

Luo, S. "Introduction to Recommender Systems." *Towards Data Science*, 10 December 2018. https://towardsdatascience.com/intro-to-recommender-system-collaborative-filtering-64a238194a26.

MacAvaney, S., H-R. Yao, E. Yang, K. Russell, N. Goharian and O. Frieder. "Hate speech detection: Challenges and solutions." *PLoS ONE* 14, issue 8 (August 2019): e0221152. DOI: 10.1371/journal.pone.0221152.

MacManus, D., K. Dean, A. C. Iversen, L. Hull, N. Jones, T. Fahy, S. Wessely, N. T. Fear. "Impact of pre-enlistment antisocial behaviour on behavioural outcomes among UK military personnel." *Social Psychiatry and Psychiatric Epidemiology* 47 (2012): 1353-1358. DOI:10.1007/s00127-011-0443-z.

MacManus, D., K. Dean, M. Jones, R. J. Rona, N. Greenberg, L. Hull, T. Fahy, S. Wessely and N. T. Fear. "Violent offending by UK military personnel deployed to Iraq and Afghanistan: a data linkage cohort study." *The Lancet* 381 (April 2013): 907-917.

Mercer, D. "Coronavirus: Hate crimes against Chinese people soar in UK during COVID-19 crisis." *Sky News*, 5 May 2020. https://news.sky.com/story/coronavirus-hate-crimes-against-chinese-people-soar-in-uk-during-covid-19-crisis-11979388.

Messer, S. C., X. Liu, C. W. Hoge, D. N. Cowan and C. C. Engel Jnr. "Projecting mental disorder prevalence from national surveys to populations-of-interest: An illustration using ECA data and the U.S. Army." *Social Psychiatry and Psychiatric Epidemiology* 39, issue 6 (June 2004): 419-426. DOI:10.1007/s00127-004-0757-1.

Moore, M. and G. Ramsay. "UK media coverage of the 2016 EU Referendum campaign." King's College London. Centre for the Study of Media, Communication and Power. May 2017. https://www.kcl.ac.uk/policy-institute/assets/cmcp/uk-media-coverage-of-the-2016-eu-referendum-campaign.pdf.

Neapolitan, R. E. and X. Jiang. *Artificial Intelligence: With an Introduction to Machine Learning*. Second Edition. Boca Raton, FL: CRC Press, 2018.

Nielsen, M. A. "Neural Networks and Deep Learning." *Determination Press*. Last updated 26 December 2019. http://neuralnetworksanddeeplearning.com/.

O'Connell, M. E. "Banning Autonomous Killing: The Legal and Ethical Requirement that Humans Make Near-Time Lethal Decisions." In *The American Way of Bombing: Changing Ethical and Legal Norms from Flying Fortresses to Drones*, edited by Matthew Evangelista and Henry Shue, 224-236. Ithaca: Cornell University Press, 2014.

Patel, T. G., and L. Connelly. "'Post-race' racisms in the narratives of 'Brexit' voters." *The Sociological Review* 67, issue 5 (September 2019): 968-984.

Paulhus, D. L. and K. M. Williams. "The Dark Triad of personality: Narcissism, Machiavellianism, and psychopathy." *Journal of Research in Personality* 36, issue 6 (December 2002): 556-563.

Pew Research Center. "Being Christian in Western Europe." Religion and Public Life, 29 May 2018. https://www.pewforum.org/2018/05/29/being-christian-in-western-europe/.

Pew Research Center. "In the U.S. and Western Europe, people say they accept Muslims, but opinions are divided on Islam." Fact Tank News in Numbers, 8 October 2019. https://www.pewresearch.org/fact-tank/2019/10/08/in-the-u-s-and-western-europe-people-say-they-accept-muslims-but-opinions-are-divided-on-islam/.

Piazza, J. A. "Politician hate speech and domestic terrorism." *International Interactions* (March 2020). DOI: 10.1080/03050629.2020.1739033.

Schulzke, M. "Ethically Insoluble Dilemmas in War." *Journal of Military Ethics* 12, no. 2 (July 2013): 95-110. DOI: 10.1080/15027570.2013.818406.

Sharma, P. "Build Your First Multi-Label Image Classification Model in Python." Analytics Vidhya, 15 April 2019. https://www.analyticsvidhya.com/blog/2019/04/build-first-multi-label-image-classification-model-python/.

Stacey, K. and J. Pickard. "Coronavirus pandemic boosts popularity of Trump and Johnson." *Financial Times*. https://www.ft.com/content/c7f5a8bc-eb0e-45e5-a080-bbfd6d317def.

United States Senate. "Putin's Asymmetric Assault on Democracy in Russia and Europe: Implications for U.S. National Security." Committee on Foreign Relations, 10 January 2018. https://www.foreign.senate.gov/imo/media/doc/FinalRR.pdf.

Volkert, J., T-C. Gablonski and S. Rabung. "Prevalence of personality disorders in the general adult population in Western countries: systematic review and meta-analysis." *The British Journal of Psychiatry* 213, issue 6 (December 2018): 709-715. DOI: 10.1192/bjp.2018.202.

Warner, C. H., G. N. Appenzeller, A. Mobbs, J. R. Parker, C. M. Warner, T. Grieger, C. W. Hoge. "Effectiveness of Battlefield-Ethics Training During Combat Deployment: A Programme Assessment." *The Lancet* 378, issue 9794 (September 2011): 915-924.

Whitley, B. E. "Right-Wing Authoritarianism, Social Dominance Orientation, and Prejudice." *Journal of Personality and Social Psychology* 77, no. 1 (1999): 126-134.

World Health Organisation. "Social Stigma associated with COVID-19." 24 February 2020. https://www.who.int/docs/default-source/coronaviruse/covid19-stigma-guide.pdf.

Yong M., T. Jiang, C. Jin, Q. Li and X. Jin. "Endogenetic structure of filter bubble in social networks." *Royal Society Open Science* 6, no. 11 (November 2019): 190868.

10

AI ETHICS

Scott Robbins

Introduction

Artificial intelligence (AI) methodologies have improved dramatically over the last decade. The field of machine learning has made advances in AI that many thought not possible. It is predicted that AI will transform every sector of society—including intelligence, surveillance, and counter-terrorism. There are, however, many pitfalls to using modern AI technologies. Ignoring them could not only violate human rights and critical ethical principles but hinder rather than enhance government operations.

In this chapter, two case studies are presented to highlight the major ethical issues surrounding the state's use of AI. The first case study examines the case of someone denied the ability to travel to the United States (U.S.) because an AI algorithm classified them as a security threat. This case study highlights AI's major ethical issues like biased training data, disparate impact, algorithmic opacity, and meaningful human control. AI concepts like precision and recall, machine learning, and explainable AI will also be discussed.

The second case study explores the quest to establish meaningful human control over autonomous weapons. As these weapons are powered by opaque artificial intelligence, having a human being with the requisite knowledge and control over the machine's outputs is seen as a significant problem. Proposed methods for establishing such control will be explained—including machine ethics (programming AI to be guided by ethical principles), human on the loop, human in the loop, and track and trace. Readers will become familiar with the difficult debate surrounding meaningful human control and the proposals to solve it.

Finally, I will point towards my solution to the many ethical problems highlighted in this chapter.

Being Denied Entry to the U.S. by an Algorithm

On 12 February 2020, the United States Embassy in London notified Eyal Weizman that his ESTA (visa waiver) had been revoked. Therefore, he would not be allowed to board a previously scheduled flight to Miami on 14 February. His wife and children, who were flying ahead of him, were stopped at JFK airport in New York. His children were separated from his wife, who was detained and questioned for two and a half hours before being allowed entry.[1] Eyal Weizman was flagged by an algorithm developed by a Virginia based firm called DataRobot.[2] DataRobot was contracted by the U.S. Department of Homeland Security (DHS) to develop "predictive models to enhance identification of high-risk passengers" in software that should "make real-time prediction[s] with a reasonable response time."[3]

The analyst who received Eyal Weizman at the U.S. embassy in London explained that "the algorithm" had identified him as a security threat. Unfortunately, the analyst could not explain how the algorithm came to that conclusion. The analyst required 15 years of travel history and the names of anyone in his network that may have triggered the algorithm. Weizman declined to give that information.

This case study highlights many of the ethical issues surrounding the use of AI in a security context. The very first issue which must be discussed is that of *algorithmic opacity*. Algorithmic opacity points to the fact that contemporary AI methodologies, specifically those falling under the umbrella of *Machine Learning* (ML), cannot explain how they reach a particular output. In effect, the algorithm is a black box—incomprehensible to human beings. We know the input and output; however, the considerations that played a factor in leading to that output are unknown to us.

ML is the methodology driving much of the excitement surrounding AI today. Rather than explicitly coding rules for a program to follow, ML uses statistical methods to "learn" how to, for example, classify images. We feed the algorithm with many pre-labeled images (for example, of guns).

---

1 Eyal Weizman, "I Was Denied Entry into the U.S. Because of a 'Homeland Security Algorithm,'" Fast Company, February 20, 2020, https://www.fastcompany.com/90466400/i-was-denied-entry-into-the-u-s-im-not-a-terrorist-i-investigate-human-rights-abuses.
2 Robert Mackey, "Homeland Security Algorithm Revokes U.S. Visa of War Crimes Investigator Eyal Weizman," The Intercept, February 21, 2020, https://theintercept.com/2020/02/20/homeland-security-algorithm-revokes-u-s-visa-war-crimes-investigator-eyal-weizman/.
3 Sam Biddle, "Homeland Security Will Let Computers Predict Who Might Be a Terrorist on Your Plane — Just Don't Ask How It Works," The Intercept, December 3, 2018, https://theintercept.com/2018/12/03/air-travel-surveillance-homeland-security/.

After thousands of images, the program can take novel images that have not yet been labeled and classify them as having a gun or not.[4] Algorithms trained this way are pretty effective; however, when they do get it wrong, they are *really* wrong. For example, in an infamous case in 2015, Google's image classification system (which powers Google image search) classified a black woman as a gorilla.[5]

Furthermore, these algorithms can be fooled. Researchers recently showed that they could trick an algorithm into thinking a turtle is a gun.[6] This is because the considerations leading to a potential output are not restricted to ones that are human articulable. By expanding the space of reasoning beyond what could be comprehended by a human being, an algorithm is able to do things that humans cannot. Algorithmic opacity is not just a problem of ML; instead, it is the reason ML is so powerful. There is no way an algorithm programmed with specific rules could beat the world champion GO player Lee Sodol. However, an algorithm powered by ML was able to do so multiple times.[7]

Algorithmic opacity, however, leads to many problems of ethical import. The first is that those affected by these algorithms' outputs do not know why they were labeled in a certain way. Weizman does not know why he was labeled a security threat. This prevents Weizman and others from being able to exercise any recourse over the algorithm. That is, they cannot challenge the algorithm's output. Suppose there was an explanation claiming, for example, that Weizman had frequently traveled to Pakistan. In that case, Weizman could perhaps challenge the algorithm by showing he had not traveled to Pakistan.

Furthermore, the U.S. embassy analyst does not have any information regarding how the algorithm came to label Weizman a security threat. This makes it incredibly difficult for the analyst to check whether or not Weizman is, in fact, a security threat. If, for example, Weizman shares his name with

---

4 See for example, Gyanendra K. Verma and Anamika Dhillon, "A Handheld Gun Detection Using Faster R-CNN Deep Learning," in *Proceedings of the 7th International Conference on Computer and Communication Technology*, ICCCT-2017 (New York, NY, USA: ACM, 2017), 84–88, https://doi.org/10.1145/3154979.3154988.
5 BBC News, "Google Apologises for Photos App's Racist Blunder," *BBC News*, July 1, 2015, sec. Technology, http://www.bbc.com/news/technology-33347866.
6 Alex Hern, "Shotgun Shell: Google's AI Thinks This Turtle Is a Rifle," *The Guardian*, November 3, 2017, sec. Technology, https://www.theguardian.com/technology/2017/nov/03/googles-ai-turtle-rifle-mit-research-artificial-intelligence.
7 Agence France-Presse, "World's Best Go Player Flummoxed by Google's 'Godlike' AlphaGo AI," *The Guardian*, May 23, 2017, sec. Technology, https://www.theguardian.com/technology/2017/may/23/alphago-google-ai-beats-ke-jie-china-go.

a known terrorist who is currently in jail in France, and that is why the algorithm considers him as a security threat, then the analyst could quickly understand that this Weizman, sitting in front of him in London, is not a security threat. However, in this case, Weizman is prevented from traveling and must endure intrusive questioning without any articulable cause. That is, there is no justification other than the output of the algorithm for this denial of travel and intrusive questioning.

Algorithmic opacity has been a problem in many contexts, including: judicial sentencing—where algorithms determine which convicted criminals deserve higher prison sentences; predictive policing—where certain people are flagged as likely to commit a crime; loan application rejection/acceptance; etc. In each of these areas, it appears that a justifying explanation must be provided for the people affected to be subjected to the consequences of how this algorithm labels you. If you get a higher prison sentence than someone else who committed the same crime, it seems appropriate that you should get an answer as to why that is the case. Weizman also deserves an answer. Various methods are being worked on to try and make so-called "explainable AI." While there is some exciting work being done, there is no solution to this problem that overcomes the ethical issues associated with algorithmic opacity.[8]

One thing we can look to as a source of possible problems causing false-positives and false-negatives in algorithmic outputs is how that algorithm was built. We need to specifically consider how that algorithm was trained. As I previously noted above, ML algorithms use massive datasets that are pre-labeled to learn how to classify future inputs. Two values are essential when designing a classification algorithm: "precision" and "recall." Precision is the total number of true-positives divided by the number of true + false positives. Suppose an algorithm is classifying people as terrorists. In that case, the precision rate gives us the percentage when the algorithm classifies someone as a terrorist, they are indeed a terrorist. If you have a low precision rate, then there is a low chance that when the algorithm classifies someone as a terrorist that they are indeed a terrorist. Recall is the total number of true positives divided by the total number of true positives + the total number of false negatives. So if there are 1000 total terrorists and the algorithm classifies

---

[8] Scott Robbins, "A Misdirected Principle with a Catch: Explicability for AI," *Minds and Machines* 29, no. 4 (December 1, 2019): 495–514, https://doi.org/10.1007/s11023-019-09509-3; Scott Robbins, "Machine Learning & Counter-Terrorism: Ethics, Efficacy, and Meaningful Human Control" (Doctoral Thesis, Delft, The Netherlands, Technical University of Delft, 2021), https://repository.tudelft.nl/islandora/object/uuid:ad561ffb-3b28-47b3-b645-448771eddaff.

500 of them as terrorists, then the recall rate is 50 percent. Both of these values are really important; however, they are in tension with one another. An everyday example may help illustrate. While walking around campus I come across many people. Some will be my students, colleagues, friends, etc. I would like to smile and wave at the people I know. However, I would be embarrassed if I were to smile and wave at someone I do not know (not to mention the way it could be interpreted by that person). Therefore, I would like to be precise with my smiles and waves—that is, when I smile and wave to someone, that person is actually someone I know. This, in turn, reduces my recall rate—that is, because I am being cautious about this (and increasing my precision) I am likely to not smile and wave at some people I know.[9]

In a terrorism context, if you maximize precision, then you are likely to lower your recall and vice versa. For example, you can achieve a 100 percent recall rate by simply labeling everyone a terrorist. If your algorithm is incredibly precise, then you are likely to miss many terrorists. Social media companies have a high-profile problem dealing with these two conflicting values as they try to take down terrorist propaganda and hate speech without taking down many false-positives (for example, newspaper articles depicting terrorists).[10] Weizman may simply be a victim of recall being favored over precision. Someone may have decided to design the algorithm in such a way as to flag a wide variety of people to capture as many terrorists as possible. This, as we have seen, increases the likelihood that non-terrorists are labeled as terrorists.

Even if we solve the precision/recall dilemma, an algorithm could be trained with *biased training data*. To illustrate what biased training data is, we can look at a simple example of Amazon using an algorithm to decide whom to hire. After some time using the algorithm, people noticed that only men were being hired for management positions. It turns out that the data Amazon used to train its algorithm was the data associated with hiring it had done in the past. It just so happens that past hiring practices were significantly biased towards men. The algorithm learns these biases through the training data and then spits them right back out in practice (for example, garbage-in/garbage-out).

---

9 Thanks to Michael Skerker for this example.
10 Steven Overly, "Facebook Plans to Use AI to Identify Terrorist Propaganda," *Washington Post*, February 16, 2017, sec. Innovations, https://www.washingtonpost.com/news/innovations/wp/2017/02/16/facebook-plans-to-use-ai-to-identify-terrorist-propaganda/; Rachel Kaser, "Google Upgrades AI to Flag Propaganda Videos," The Next Web, April 3, 2017, https://thenextweb.com/google/2017/04/03/google-upgrades-ai-flag-propaganda-vids/.

In a counter-terrorism context, the worry here is that the training data might also be biased. Because past terrorist data might be highly skewed towards Muslims, the algorithm might merely be picking out Muslims—or those with names that are "weird." This causes two problems. First, someone might be subjected to increased scrutiny solely based on their membership in a group; that is, the group in question may be *disparately impacted* by the algorithm. Disparate impact is the idea that the benefits and harms of a particular algorithm are unequally distributed among groups. So while middle-aged white men might benefit from the speedy security screening process at airports, Muslims and people of color not only will not feel those benefits, but will unequally be subjected to the harm of having an increased likelihood of being a false-positive of the algorithm. Second, the algorithm is likely to miss new threats because its data is biased towards other groups. For example, far-right terrorists may be missed because the training data did not include data about them.

An excellent example of biased training data leading to disparate impact is that of facial recognition algorithms. Joy Buolamwini and Timnit Gebru showed that facial recognition algorithms had a 34 percent failure rate for women of color while only a 0.8 percent failure rate for white men.[11] This means that black women have a 34 percent chance of either not being identified at all (making it hard to get past security in airports) or being misidentified. This disproportionately impacts this group for no other reason than their gender and skin color. This brings us back to Eyal Weizman. Maybe he is part of a group that is disparately impacted by the algorithm.

Whatever the reason, something went wrong for him. Eyal Weizman is a professor at Goldsmiths University in London. He runs a research agency called Forensic Architecture that investigates war crimes. Weizmann was traveling to the U.S. to showcase the work that his research agency had done. In investigating war crimes, he has many contacts with victims of those war crimes. He could not give specific information about the people he was in contact with because of their vulnerability (some of the possible war crimes were perpetrated by the United States). The lack of explanation prevents him, or anyone else, from pursuing *recourse*. That is, they are unable to acquire the information they need to challenge the decision—whether it is based on wrong information (maybe the algorithm believes

---

11 Joy Buolamwini and Timnit Gebru, "Gender Shades: Intersectional Accuracy Disparities in Commercial Gender Classification," in *Conference on Fairness, Accountability and Transparency,* 2018, 77–91, http://proceedings.mlr.press/v81/buolamwini18a.html.

he traveled to Syria when he did not) or because the explanation doesn't justify the decision. Some academics are proposing methods and principles to ensure that those labeled by such algorithms can pursue recourse.[12]

Killer Robots and Meaningful Human Control

The pursuit of increasingly autonomous weapons is real. Automating the targeting and firing of a weapon, it is argued, could allow for increased speed and accuracy—and keep soldiers out of the line of fire. Furthermore, some have argued that because machines lack emotions and fatigue, they will better be able to follow the rules of engagement and never have the motivation to rape, torture, or murder innocent civilians.[13]

However, according to the U.S. Department of Defense (DoD) and many others, decisions of life and death must remain under *meaningful human control*. In Directive 3000.09, the U.S. DoD states that "autonomous and semi-autonomous weapon systems shall be designed to allow commanders and operators to exercise appropriate levels of human judgment over the use of force." Human Rights Watch argues that:

> Mandating meaningful human control of weapons would help protect human dignity in war, ensure compliance with international humanitarian and human rights law, and avoid creating an accountability gap for the unlawful acts of a weapon. [14]

Now philosophers, lawyers, and policy-makers are trying to figure out what meaningful human control amounts to. Specifically, what makes human control *meaningful*?

Let's imagine an autonomous drone that uses facial recognition to

---

12 see for example, Berk Ustun, Alexander Spangher, and Yang Liu, "Actionable Recourse in Linear Classification," in *Proceedings of the Conference on Fairness, Accountability, and Transparency*, FAT* '19 (New York, NY, USA: ACM, 2019), 10–19, https://doi.org/10.1145/3287560.3287566.
13 Ronald Arkin, Patrick Ulam, and Alan R. Wagner, "Moral Decision Making in Autonomous Systems: Enforcement, Moral Emotions, Dignity, Trust, and Deception," *Proceedings of the IEEE* 100, no. 3 (March 2012): 571–89, https://doi.org/10.1109/JPROC.2011.2173265; Ronald Arkin, "Governing Lethal Behavior: Embedding Ethics in a Hybrid Deliberative/Reactive Robot Architecture," in *Proceedings of the 3rd ACM/IEEE International Conference on Human Robot Interaction*, HRI '08 (New York, NY, USA: Association for Computing Machinery, 2008), 121–28, https://doi.org/10.1145/1349822.1349839.
14 "Killer Robots and the Concept of Meaningful Human Control," Human Rights Watch, April 11, 2016, https://www.hrw.org/news/2016/04/11/killer-robots-and-concept-meaningful-human-control.

find previously known terrorists to explore this topic. If the drone finds one of these terrorists, it can autonomously fire its weapon to terminate that terrorist provided no civilians will be killed.[15] As described, there is no human control at all. Because this violates DODD 3000.09, it is decided that a human being must be "on-the-loop." *Human-on-the-loop* describes a situation in which a human can watch an entirely autonomous algorithm. A human being has the power to stop the algorithm if she decides that such an algorithm is incorrect or doing the wrong thing.

In the example we are working with, a human being could decide that although the algorithm has correctly identified a terrorist who should indeed be killed, there would be too many civilian deaths to go ahead with the kill. In this way, it is said that this human being would be in control of the algorithm.

However, this suffers from several problems that prevent this control from being *meaningful*. An individual would not have time to gain situational awareness to make an informed decision on the matter. Furthermore, human beings suffer from biases that will be difficult to overcome in such a situation. One of these biases is *automation bias*, whereby a "human decision-maker disregards or does not search for contradictory information in light of a computer-generated solution which is accepted as correct."[16] A second bias is *confirmation bias*, which occurs when "humans seek out evidence that confirms their prior beliefs or the hypothesis on hand."[17] Not only will humans tend not to search for any evidence that contradicts the machine, but they will actively search for evidence that confirms the machine's output. It has been shown that these biases result in human beings performing better when the algorithm is correct; however, they perform much worse when the algorithm gets it wrong.[18]

These biases make the control that the human on the loop has over this autonomous weapon not very meaningful. No human being should accept

---

15 Some like this is described as the most likely version of an autonomous weapons system in Kelsey Piper, "Death by Algorithm: The Age of Killer Robots Is Closer than You Think," Vox, June 21, 2019, https://www.vox.com/2019/6/21/18691459/killer-robots-lethal-autonomous-weapons-ai-war.
16 Mary Cummings, "Automation Bias in Intelligent Time Critical Decision Support Systems," in *AIAA 1st Intelligent Systems Technical Conference* (American Institute of Aeronautics and Astronautics, 2012), https://doi.org/10.2514/6.2004-6313.
17 Robbins, "Machine Learning & Counter-Terrorism: Ethics, Efficacy, and Meaningful Human Control," 10.
18 Linda J. Skitka, Kathleen L. Mosier, and Mark Burdick, "Does Automation Bias Decision-Making?," *International Journal of Human-Computer Studies* 51, no. 5 (November 1, 1999): 991–1006, https://doi.org/10.1006/ijhc.1999.0252.

responsibility for the consequences that this algorithm will have simply because they have the power to stop the algorithm. The same problems occur for *human-in-the-loop* solutions to meaningful human control. When humans are in the loop, they must do something for the algorithm's output to cause actions. In the drone example above, the human being would have to "agree" with the algorithm before a strike occurred. Automation bias and confirmation bias are still present. Furthermore, the time-sensitive nature of a strike may not give a human enough time to gain the awareness needed to make an informed decision.

This has been studied in driverless cars. It has been shown, for example, that drivers need eight seconds to gain situational awareness of an automated car. This is too long. A crash happens in seconds. So the idea that Tesla, for example, can put humans in control by forcing them to have their hands on the wheel is untenable. When bad things happen, the human brain is not equipped to take over in time to prevent a crash.[19] The same could be true for drones.

Knowing about these biases and delays in gaining situational awareness should cause us to rethink how we have humans in control over these algorithms. The humans that we put in control must be set up to succeed. They must have the power and the knowledge to exercise the control they are given. If these biases and delays in gaining situational awareness prevent such power and knowledge, we have not achieved meaningful human control.

Some academic researchers and governments have put forward an entirely different solution to meaningful human control. Rather than have humans in control over ethical decisions, it is argued that machines could be endowed with ethical reasoning capabilities to make these decisions themselves. This is called *machine ethics*. For example, it has been argued by Professor Ron Arkin that machines could have the Laws of War (LOW) and the rules of engagement (ROE) embedded into them. This would be better, according to Prof. Arkin, because machines don't get tired or angry. They would have no reason to inflict retributive attacks on innocent populations. In short, there would be no Mai Lai massacre or Abu Ghraib if machines

---

19 Zhenji Lu, Xander Coster, and Joost de Winter, "How Much Time Do Drivers Need to Obtain Situation Awareness? A Laboratory-Based Study of Automated Driving," *Applied Ergonomics* 60 (April 1, 2017): 293–304, https://doi.org/10.1016/j.apergo.2016.12.003; Timothy J. Wright et al., "Effects of Alert Cue Specificity on Situation Awareness in Transfer of Control in Level 3 Automation," *Transportation Research Record: Journal of the Transportation Research Board* 2663 (January 1, 2017): 27–33, https://doi.org/10.3141/2663-04.

were in control. Arkin says that he is "convinced that [machines] can perform more ethically than human soldiers are capable of."[20]

Millions of dollars are now funding research programs to achieve the aims of machine ethics.[21] To be sure, this is not a solution that achieves meaningful human control at all. However, it is an attempt to solve the problem by saying if machines had ethics, there would be no need for humans to be in control. In this way, the problem is re-framed. One side asks the question, "algorithms will be making important decisions; therefore, how can we maintain meaningful human control?" The other side asks, "algorithms will be making important decisions; therefore, how can we ensure that they *ethically* make these decisions?"

The idea of machine ethics faces many practical challenges. Those familiar with the Just War Theory principles, for example, know how difficult it is for us to come to conclusions about proportionality and discrimination. If we can't agree on these solutions, how can we be sure an algorithm will be doing better? Furthermore, these machines are limited by the sensors feeding them information. A sensor that fails to recognize 20 civilians because they are wearing clothes that blend in with the ground cannot make a proper ethical decision because it does not have the correct information. We know that although AI recognition systems are incredibly powerful, when they get it wrong, they get it *really* wrong. In the previous section, I mentioned some examples of these errors. There are also many examples of "adversarial examples," which are designed to trick AI algorithms.[22] If we cannot rely on the sensors feeding information to these algorithms, then it won't matter what "ethical reasoning" they have hardwired into their systems.

Machine ethics also faces many conceptual challenges. The most discussed is the so-called "responsibility gap" or "accountability gap." The idea is that when an algorithm makes an ethical decision, there will be a gap in the ascription of responsibility for that decision. If the drone in the example above autonomously decided to target and kill a terrorist, but in the end, killed 100 innocent civilians at a wedding party, who is responsible? The designers of the algorithm could not have foreseen this specific example. This input was entirely novel. No human being appears to be able

---

20 Arkin, "Governing Lethal Behavior."
21 see e.g. George Nott, "Killer Robot Campaign Defector to 'embed Ethics' in Autonomous Weapons," Computerworld, March 10, 2019, https://www.computerworld.com/article/3457068/killer-robot-campaign-defector-to-embed-ethics-in-autonomous-weapons.html.
22 Battista Biggio and Fabio Roli, "Wild Patterns: Ten Years after the Rise of Adversarial Machine Learning," *Pattern Recognition* 84 (December 1, 2018): 317–31, https://doi.org/10.1016/j.patcog.2018.07.023.

to receive blame or moral responsibility for the terrible outcome. Telling the families of the 100 people killed that the machine itself is to blame will not satisfy them. Machine learning makes the allocation of responsibility, accountability, and retribution complicated—if not impossible.[23] Solving this complicated problem will be necessary if machine ethics is to be a viable solution.

Other academics (including myself) have argued that: there is no upside to ethical reasoning in machines;[24] that ethical machines will hurt humans ability to ethically reason;[25] that the use of the words "ethical" and "moral" are *really* referring to "safety;" that we should not place robots in roles requiring moral reasoning;[26] and that reducing ethics to "moral calculation" limits the breadth and scope of moral concepts used in moral deliberation.[27]

In another attempt at solving the problem of meaningful human control scholars have suggested that two principles must be met by AI systems: a *tracking* condition and a *tracing* condition. The tracking condition states that a machine's decisions must be responsive to human moral considerations. For example, if the human who would be in charge of the drone were not to initiate a strike because a child was playing soccer nearby, the drone should also not initiate a strike if it sees a child playing soccer nearby. Considerations that would play a role in a human's moral deliberation should also play a role in the machine's outcome.

The second condition states that the machine's "actions/states should

---

[23] For those interested in this discussion see Robert Sparrow, "Killer Robots," *Journal of Applied Philosophy* 24, no. 1 (2007): 62–77, https://doi.org/10.1111/j.1468-5930.2007.00346.x; John Danaher, "Robots, Law and the Retribution Gap," *Ethics and Information Technology* 18, no. 4 (December 1, 2016): 299–309, https://doi.org/10.1007/s10676-016-9403-3; Sven Nyholm, "Attributing Agency to Automated Systems: Reflections on Human–Robot Collaborations and Responsibility-Loci," *Science and Engineering Ethics* 24, no. 4 (August 1, 2018): 1201–19, https://doi.org/10.1007/s11948-017-9943-x; Roos de Jong, "The Retribution-Gap and Responsibility-Loci Related to Robots and Automated Technologies: A Reply to Nyholm," *Science and Engineering Ethics* 26, no. 2 (April 1, 2020): 727–35, https://doi.org/10.1007/s11948-019-00120-4.
[24] Aimee van Wynsberghe and Scott Robbins, "Critiquing the Reasons for Making Artificial Moral Agents," *Science and Engineering Ethics* 25, no. 3 (June 1, 2019): 719–35, https://doi.org/10.1007/s11948-018-0030-8.
[25] Shannon Vallor, "Moral Deskilling and Upskilling in a New Machine Age: Reflections on the Ambiguous Future of Character," *Philosophy & Technology* 28, no. 1 (March 1, 2015): 107–24, https://doi.org/10.1007/s13347-014-0156-9.
[26] Amanda Sharkey, "Can We Program or Train Robots to Be Good?," *Ethics and Information Technology* 22, no. 4 (December 1, 2020): 283–95, https://doi.org/10.1007/s10676-017-9425-5.
[27] Christian Herzog, "Three Risks That Caution Against a Premature Implementation of Artificial Moral Agents for Practical and Economical Use," *Science and Engineering Ethics* 27, no. 1 (January 26, 2021): 3, https://doi.org/10.1007/s11948-021-00283-z.

be traceable to a proper moral understanding on the part of one or more relevant human persons who design or interact with the system."[28] That is, there must be a human being who knows the capabilities of the machine and the understanding that they can receive legitimate moral blame for the consequences of that machine. These two conditions are a significant step toward showing what is required by meaningful human control. However, the difficulty lies in achieving these conditions.

Artificial General Intelligence

A word must be said about the concern of *superintelligence* or *artificial general intelligence*. Some notable technologists and academics have expressed concerns that AI could become so advanced that it could surpass human-level intelligence.[29] Rather than only being able to do one thing (for example, classify images), these machines could solve any problem given to it. That is, they would have *general* intelligence rather than *narrow*. These machines, it is claimed, could decide that the best way to protect the state is to, for example, destroy everyone else in the world.[30] The super-intelligent machine could find novel shortcuts that humans find abhorrent to solve problems. In this case, eliminate all the risks outside of the state.

While this is a fascinating thought experiment and fuel for science fiction books and movies, we are no closer to this level of intelligence than we were 50 years ago. The philosopher Luciano Floridi describes the progress we have made as analogous to saying you are closer to getting to

---

28 Filippo Santoni de Sio and Jeroen van den Hoven, "Meaningful Human Control over Autonomous Systems: A Philosophical Account," *Frontiers in Robotics and AI* 5 (2018), https://doi.org/10.3389/frobt.2018.00015.
29 see e.g. Matt McFarland, "Elon Musk: 'With Artificial Intelligence We Are Summoning the Demon.,'" Washington Post, October 24, 2014, https://www.washingtonpost.com/news/innovations/wp/2014/10/24/elon-musk-with-artificial-intelligence-we-are-summoning-the-demon/; Rory Cellan-Jones, "Stephen Hawking Warns Artificial Intelligence Could End Mankind," *BBC News*, December 2, 2014, sec. Technology, http://www.bbc.com/news/technology-30290540.
30 Nick Bostrom and Eliezer Yudkowsky, "The Ethics of Artificial Intelligence," *The Cambridge Handbook of Artificial Intelligence*, 2014, 316–34; Nick Bostrom, "Existential Risks: Analyzing Human Extinction Scenarios," *Journal of Evolution and Technology* 9, no. 1 (2001): 1–31; Anthony M. Barrett and Seth D. Baum, "Risk Analysis and Risk Management for the Artificial Superintelligence Research and Development Process," in *The Technological Singularity* (Springer, 2017), 127–40.

the moon because you have learned to climb a tree.³¹ There is simply no good reason to believe that anything like superintelligence will exist in the foreseeable future. Most importantly, this debate distracts from the genuine concerns posed by the technology we have right now.

Conclusion

The many ethical issues associated with AI discussed above should not suggest that the state should not use AI. Knowledge of these challenges should help us understand the contexts in which AI could be most helpful without falling victim to one of the above challenges. In my work, I argue that meaningful human control is about controlling where these algorithms are used more than it is about controlling their individual outputs. We can use the simple example of a chainsaw to illustrate my approach. Rather than figuring out how to solve the safety issues of chainsaws being used in the context of a daycare center, my approach argues that we should not be using chainsaws in daycare centers. We simply know enough about the dangers of chainsaws to know that it would be unethical to use them around a bunch of small children.

Knowing that AI has problems of bias, disparate impact, and unexplainable decisions should cause us to look for contexts to use AI where these won't cause ethical problems. Algorithms used to detect weapons in airport security bag scans do not fall victim to any of the problems above. There are no people involved, so bias and disparate impact are not relevant. And we simply don't care how the algorithm comes to the decision that there is a gun inside the bag. We only care if the algorithm was correct or not.

The first step towards being able to have meaningful control over algorithms is to understand enough about these algorithms to decide what they can and should be doing. I argue that we must know information about the training data, possible inputs, possible outputs, functions (or aims), and boundaries of the algorithm.³² Knowing about the training data, for example, helps a human being have the knowledge necessary

---

31 Luciano Floridi, "True AI Is Both Logically Possible and Utterly Implausible – Luciano Floridi | Aeon Essays," Aeon, May 9, 2016, https://aeon.co/essays/true-ai-is-both-logically-possible-and-utterly-implausible.
32 Scott Robbins, "AI and the Path to Envelopment: Knowledge as a First Step towards the Responsible Regulation and Use of AI-Powered Machines," *AI & SOCIETY* 35, no. 2 (June 1, 2020): 391–400, https://doi.org/10.1007/s00146-019-00891-1.

to understand whether this particular algorithm will be unacceptably biased. Suppose an algorithm designed to detect suspicious behaviour, for example, was only trained on Westerners. In that case, it is entirely likely that the novel behavior of people from other cultures will be labeled as suspicious because the machine has not encountered that behavior in its training. Therefore, a human being is in a position to understand the algorithm enough for the consequences of that algorithm to be "traced" back to them.

Moral responsibility should not be delegated to a human being for a particular output of the machine; instead, it is the moral responsibility of placing a machine into a particular context that merits moral responsibility. I have already committed a grave ethical offense if I place my 7-year-old daughter in charge of watching her baby brother in the bath. The same can be said of placing an algorithm where it is known that it will have the problems highlighted above, or, due to lack of diligence, not knowing whether the algorithm has these problems.

Artificial intelligence is not a silver bullet that will solve all of our problems. It is a potent tool that comes with limitations—limitations that are so far poorly understood by many who use it. When understood, these limitations put human beings in a position to harness the power of AI without suffering the many ethical pitfalls discussed above.

References

Agence France-Presse, "World's Best Go Player Flummoxed by Google's 'Godlike' AlphaGo AI," *The Guardian*, May 23, 2017, sec. Technology, https://www.theguardian.com/technology/2017/may/23/alphago-google-ai-beats-ke-jie-china-go.

Arkin, R., P. Ulam, and A. R. Wagner, "Moral Decision Making in Autonomous Systems: Enforcement, Moral Emotions, Dignity, Trust, and Deception," *Proceedings of the IEEE* 100, no. 3 (March 2012): 571–89, https://doi.org/10.1109/JPROC.2011.2173265.

Arkin, R. "Governing Lethal Behavior: Embedding Ethics in a Hybrid Deliberative/Reactive Robot Architecture," in *Proceedings of the 3rd ACM/IEEE International Conference on Human Robot Interaction*, HRI '08 (New York, NY, USA: Association for Computing Machinery, 2008), 121–28, https://doi.org/10.1145/1349822.1349839.

Barrett, A.M. and S. D. Baum, "Risk Analysis and Risk Management for the Artificial Superintelligence Research and Development Process," in *The Technological Singularity* (Springer, 2017), 127–40.

BBC News, "Google Apologises for Photos App's Racist Blunder," *BBC News*, July 1, 2015, sec. Technology, http://www.bbc.com/news/technology-33347866.

Biddle, S. "Homeland Security Will Let Computers Predict Who Might Be a Terrorist on Your Plane — Just Don't Ask How It Works," The Intercept, December 3, 2018, https://theintercept.com/2018/12/03/air-travel-surveillance-homeland-security/.

Biggio, B. and F. Roli, "Wild Patterns: Ten Years after the Rise of Adversarial Machine Learning," *Pattern Recognition* 84 (December 1, 2018): 317–31, https://doi.org/10.1016/j.patcog.2018.07.023.

Bostrom, N. and E. Yudkowsky, "The Ethics of Artificial Intelligence," *The Cambridge Handbook of Artificial Intelligence*, 2014, 316–34; Nick Bostrom, "Existential Risks: Analyzing Human Extinction Scenarios," *Journal of Evolution and Technology* 9, no. 1 (2001): 1–31.

Buolamwini, J. and T. Gebru, "Gender Shades: Intersectional Accuracy Disparities in Commercial Gender Classification," in *Conference on Fairness, Accountability and Transparency*, 2018, 77–91, http://proceedings.mlr.press/v81/buolamwini18a.html.

Cellan-Jones, R. "Stephen Hawking Warns Artificial Intelligence Could End Mankind," *BBC News*, December 2, 2014, sec. Technology, http://www.bbc.com/news/technology-30290540.

Cummings, M. "Automation Bias in Intelligent Time Critical Decision Support Systems," in *AIAA 1st Intelligent Systems Technical Conference* (American Institute of Aeronautics and Astronautics, 2012), https://doi.org/10.2514/6.2004-6313.

Danaher, J. "Robots, Law and the Retribution Gap," *Ethics and Information Technology* 18, no. 4 (December 1, 2016): 299–309, https://doi.org/10.1007/s10676-016-9403-3.

de Jong, R. "The Retribution-Gap and Responsibility-Loci Related to Robots and Automated Technologies: A Reply to Nyholm," *Science and Engineering Ethics* 26, no. 2 (April 1, 2020): 727–35, https://doi.org/10.1007/s11948-019-00120-4.

Hern, A. "Shotgun Shell: Google's AI Thinks This Turtle Is a Rifle," *The Guardian*, November 3, 2017, sec. Technology, https://www.theguardian.com/technology/2017/nov/03/googles-ai-turtle-rifle-mit-research-artificial-intelligence.

Herzog, C. "Three Risks That Caution Against a Premature Implementation of Artificial Moral Agents for Practical and Economical Use," *Science and Engineering Ethics* 27, no. 1 (January 26, 2021): 3, https://doi.org/10.1007/s11948-021-00283-z.

Human Rights Watch. "Killer Robots and the Concept of Meaningful Human Control," April 11, 2016, https://www.hrw.org/news/2016/04/11/killer-robots-and-concept-meaningful-human-control.

Floridi, L. "True AI Is Both Logically Possible and Utterly Implausible – Luciano Floridi | Aeon Essays," Aeon, May 9, 2016, https://aeon.co/essays/true-ai-is-both-logically-possible-and-utterly-implausible.

Kaser, R. "Google Upgrades AI to Flag Propaganda Videos," The Next Web, April 3, 2017, https://thenextweb.com/google/2017/04/03/google-upgrades-ai-flag-propaganda-vids/.

Lu, Z., X. Coster, and J. de Winter, "How Much Time Do Drivers Need to Obtain Situation Awareness? A Laboratory-Based Study of Automated Driving," *Applied Ergonomics* 60 (April 1, 2017): 293–304, https://doi.org/10.1016/j.apergo.2016.12.003.

Mackey, R. "Homeland Security Algorithm Revokes U.S. Visa of War Crimes Investigator Eyal Weizman," The Intercept, February 21, 2020, https://theintercept.com/2020/02/20/homeland-security-algorithm-revokes-u-s-visa-war-crimes-investigator-eyal-weizman/.

McFarland, M. "Elon Musk: 'With Artificial Intelligence We Are Summoning the Demon.,'" Washington Post, October 24, 2014, https://www.washingtonpost.com/news/innovations/wp/2014/10/24/elon-musk-with-artificial-intelligence-we-are-summoning-the-demon/.

Nott, G. "Killer Robot Campaign Defector to 'embed Ethics' in Autonomous Weapons," Computerworld, March 10, 2019, https://www.computerworld.com/article/3457068/killer-robot-campaign-defector-to-embed-ethics-in-autonomous-weapons.html.

Nyholm, S. "Attributing Agency to Automated Systems: Reflections on Human–Robot Collaborations and Responsibility-Loci," *Science and Engineering Ethics* 24, no. 4 (August 1, 2018): 1201–19, https://doi.org/10.1007/s11948-017-9943-x.

Overly, S. "Facebook Plans to Use AI to Identify Terrorist Propaganda," *Washington Post*, February 16, 2017, sec. Innovations, https://www.washingtonpost.com/news/innovations/wp/2017/02/16/facebook-plans-to-use-ai-to-identify-terrorist-propaganda/.

Piper, K. "Death by Algorithm: The Age of Killer Robots Is Closer than You Think," Vox, June 21, 2019, https://www.vox.com/2019/6/21/18691459/killer-robots-lethal-autonomous-weapons-ai-war.

Robbins, S. "A Misdirected Principle with a Catch: Explicability for AI," *Minds and Machines* 29, no. 4 (December 1, 2019): 495–514, https://doi.org/10.1007/s11023-019-09509-3.

Robbins, S. "Machine Learning & Counter-Terrorism: Ethics, Efficacy, and Meaningful Human Control" (Doctoral Thesis, Delft, The Netherlands, Technical University of Delft, 2021), https://repository.tudelft.nl/islandora/object/uuid:ad561ffb-3b28-47b3-b645-448771eddaff.

Robbins, S. "Machine Learning & Counter-Terrorism: Ethics, Efficacy, and Meaningful Human Control," 10.

Robbins, S. "AI and the Path to Envelopment: Knowledge as a First Step towards the Responsible Regulation and Use of AI-Powered Machines," *AI & SOCIETY* 35, no. 2 (June 1, 2020): 391–400, https://doi.org/10.1007/s00146-019-00891-1.

Santoni de Sio, F. and J. van den Hoven, "Meaningful Human Control over Autonomous Systems: A Philosophical Account," *Frontiers in Robotics and AI* 5 (2018), https://doi.org/10.3389/frobt.2018.00015.

Sharkey, A. "Can We Program or Train Robots to Be Good?," *Ethics and Information Technology* 22, no. 4 (December 1, 2020): 283–95, https://doi.org/10.1007/s10676-017-9425-5.

Skitka, L.J., K. L. Mosier, and M. Burdick, "Does Automation Bias Decision-Making?," *International Journal of Human-Computer Studies* 51, no. 5 (November 1, 1999): 991–1006, https://doi.org/10.1006/ijhc.1999.0252.

Sparrow, R. "Killer Robots," *Journal of Applied Philosophy* 24, no. 1 (2007): 62–77, https://doi.org/10.1111/j.1468-5930.2007.00346.x.

Ustun, B. A. Spangher, and Y. Liu, "Actionable Recourse in Linear Classification," in *Proceedings of the Conference on Fairness, Accountability, and Transparency*, FAT* '19 (New York, NY, USA: ACM, 2019), 10–19, https://doi.org/10.1145/3287560.3287566.

Vallor, S. "Moral Deskilling and Upskilling in a New Machine Age: Reflections on the Ambiguous Future of Character," *Philosophy & Technology* 28, no. 1 (March 1, 2015): 107–24, https://doi.org/10.1007/s13347-014-0156-9.

van Wynsberghe, A. and S. Robbins, "Critiquing the Reasons for Making Artificial Moral Agents," *Science and Engineering Ethics* 25, no. 3 (June 1, 2019): 719–35, https://doi.org/10.1007/s11948-018-0030-8.

Verma, G.K. and A. Dhillon, "A Handheld Gun Detection Using Faster R-CNN Deep Learning," in *Proceedings of the 7th International Conference on Computer and Communication Technology*, ICCCT-2017 (New York, NY, USA: ACM, 2017), 84–88, https://doi.org/10.1145/3154979.3154988.

Weizman, E. "I Was Denied Entry into the U.S. Because of a 'Homeland Security Algorithm,'" Fast Company, February 20, 2020, https://www.fastcompany.com/90466400/i-was-denied-entry-into-the-u-s-im-not-a-terrorist-i-investigate-human-rights-abuses.

Wright, T.J. et al., "Effects of Alert Cue Specificity on Situation Awareness in Transfer of Control in Level 3 Automation," *Transportation Research Record: Journal of the Transportation Research Board* 2663 (January 1, 2017): 27–33, https://doi.org/10.3141/2663-04.

11

ETHICS AND CYBER SYSTEMS

Artificial Intelligent Weapons Systems and Moral Slippage

Elke Schwarz

Introduction

Innovation in automation and autonomy for military systems is advancing at a rapid speed. A growing number of countries, including the U.S., the UK, China and Russia either develop, produce and/or use military systems, including lethal ones, of varying degrees of autonomy. With developments in machine learning and computer processing power making great strides, the integration of Artificial Intelligence (AI) into military systems is likely to accelerate the shift toward full autonomy in military weapons systems substantially in the near future. The desire to develop growing levels of autonomy in military weapons technology is by no means a recent trend. Indeed, AI and military research and development programmes have traditionally co-evolved and researchers in cybernetics in the 1950s and 1960s had already raised important questions about how increased levels of machine autonomy might affect human control over military technology.

One of the early founders of cybernetic systems, Norbert Wiener, raised this issue in his 1950s text *The Human Use of Human Beings* and he warns in no uncertain terms of the risks involved in applying automated rationality to the infinitely complex and plural condition of humanity, including warfare. Wiener, having worked himself on many military applications of autonomous technologies, saw a two-fold peril in cybernetic technologies, as a mode of reasoning by a group of humans "to increase their control over the rest of the human race"[1] and as a way of outsourcing complex and difficult decisions to machine authority, "[f]or the man [...] to throw the problem of his responsibility on the machine, whether it can learn or not, is to cast his responsibility to the wind and to find it come back seated on the whirlwind."[2] An abdication of responsibility that, for him, may result in disastrous consequences. And the further the human and machine become enmeshed, the greater the risk that machine logics will win out over human sense-making, particularly so in warfare.

More than a decade of drone warfare, in which human operators have increasingly been embedded into the logic of technological systems—an assemblage of screens, data, sensors, hardware, software and humans—has given us a clear indication that the technologies we find ourselves within shape our practices, and shift goalposts, often beyond our immediate awareness.[3] While drones have proven to be useful instruments in and for warfare, there are nonetheless persistent questions about their role in alleviating the suffering of civilians in warfare.[4] To the contrary, the worries raised by critical voices, that the technical capabilities of a remotely operated lethal system might result in a lowered threshold for the use of violence appears to have materialized, at least in some contexts, as civilians often have to bear the brunt of the expanded drone wars.[5] The advent of the drone has provided the technological substrate for the autonomous weapons systems to come. Where in the early 2000s, the drone was hailed as an efficient and also ethical panacea for the problem of terrorism, today, Artificial Intelligence, and its utility for warfare, is cast in similarly hyperbolic terms. Whoever dominates the AI landscape will rule the world—to paraphrase Vladimir Putin's widely repeated claim in 2017.[6] As countries across the globe boost their resource allocation for military AI, Wiener's warning remains pertinent—we should consider where the full use of military AI might lead us if we are to avoid disastrous consequences.

---

1 Norbert Wiener, *The Human Use of Human Beings* (Boston: Houghton Mifflin, 1954), 181.
2 Wiener, *The Human Use of Human Beings*, 185.
3 Elke Schwarz, *Death Machines* (Manchester: Manchester University Press, 2019). See also Elke Schwarz, "Prescription Drones: On the Techno-biopolitical Regimes of Contemporary 'Ethical Killing'," *Security Dialogue* 47, no.1 (2016): 59-75.
4 Chris Cole, "A bloody month in the drone wars," *Drone Wars UK*, December 3, 2019.
5 Doyle McManus, "Trump's brand of war is killing more civilians than before," *Los Angeles Times*, September 8, 2019.
6 Edoardo Maggio, "Putin believes that whatever country has the best AI will be 'the ruler of the world,'" *Business Insider*, September 4, 2017.

This chapter explores the shifting goal posts for human agents in war when AI systems take on an increasing decision function, including the pre-selection and recommendation of targets, and considers the ethical challenges that arise in such a human-machine teaming. The chapter first establishes the case of AI for military use with a specific focus on applications within lethal systems, highlighting the AI use, not for fully autonomous lethal decisions, but for target recommendations, to be acted on at speed. It then traces the challenges that arise in exercising moral agency within such systems in ways that would safeguard moral decision making by human commanders and operators. Considering these challenges, the chapter suggests that the logic of the AI enabled systems in such roles is conducive to shifting weighty moral decisions onto the machine itself, thereby casting complex ethical decisions as a technical matter of utility. This side-lines complex and difficult ethical decision making, and, shifts it into the realm of problems to be solved through the technical lens of engineering, leaving little room for human operators to exercise their ethical agency for lethal outcomes.

Artificial Intelligent Weapons Systems

Conceptions of Artificial Intelligence in military domains, and specifically as a component of weapons systems (so-called Killer Robots) as depicted in popular media often invoke dystopian visions of autonomous Terminator-style monstrosities that go rogue in an inexplicable drive to destroy all humanity. The reality of AI-enabled lethal systems is considerably less spectacular, and more mundane, but, precisely for that reason, also likely more complex and challenging in moral terms. Autonomous intelligent weapons that take on a significant role in the selection and targeting functions—as they are designed and developed today by the U.S., China, Russia, and other countries—are systems that aim to increase lethality through "accelerated sensor-to-shooter timelines,"[7] which in most cases suggests that an AI systems component evaluates sensor data and makes a targeting suggestion to the operator in matters of seconds, which can then be followed through "with the click of a mouse."[8] The U.S. Department of Defenses drive and progress toward such AI weapons systems is indicative

---
7   Jack Shanahan, "Lt. Gen. Jack Shanahan media briefing on A.I.-related initiatives within the Department of Defense," August 30, 2019.
8   Nathan Strout, "Inside the Army's futuristic test of its battlefield artificial intelligence in the desert," *C4ISRNET*, September 25, 2020.

of the global trend in the AI arms race and highlights the vision for AI to play a crucial role in the accelerated identification and tracking of targets by the computer, which then leaves the human with a limited set of possible courses of action for a potentially lethal decision. At present, the U.S. DoD is partnering up with a host of private sector contractors to achieve the objective set for the immediate future: to harness the capabilities of AI for "manoeuvre and fires with individual lines of effort or product lines oriented on warfighting operations."[9] This includes the application of AI for "joint all-domain command and control, accelerated sensor-to-shooter timelines, autonomous and swarming systems" and "target development."[10] The more interaction resides with machines, it seems, the better, so as to mitigate the limited cognitive capacities of human perception and deliberation. The aim is to harness Artificial Intelligence, robotics and autonomy for a vastly accelerated kill chain in which a target can be taken out in less than 20 seconds from the capturing of targeting data to the final kill action. In practice this means linking multiple sensors with an AI enabled Command and Control (C2) node and a shooter platform, as it is, for example, the aim of the U.S. Army's *Project Convergence*. At present, the technologies are still rather immature, but the direction is clear: to leverage AI and machine learning to fire "artillery, not minutes after spotting a target, but seconds."[11] With such a sped up kill chain, the role of human decision-making at the sharp point of delivery becomes necessarily increasingly marginal.

While the DoD maintains that there are no plans underway to outsource the act of killing to the machine itself, and that this decision will continue to reside with the human, it is the explicit aim to increasingly let the AI make a *pre-selection* of possible targets at an accelerated pace. A number of specific programmes are currently in place that work toward this goal. DARPA, for example, has teamed up with Lockheed Martin for Project Squad X to deliver improved situational awareness and alleviate the "fog of war" through human-machine teaming, with "artificial intelligence as a true partner."[12] Particularly relevant here is the Mission Intelligence Tactical Systems (MITS) component, subcontracted out to BAE Systems by Lockheed Martin, which provides "advanced sensor fusion, artificial

---

9 Shanahan, "Lt. Gen. Jack Shanahan media briefing."
10 Shanahan, "Lt. Gen. Jack Shanahan media briefing."
11 Sydney Freedberg Jr., "Army's New Aim is Decision Dominance," *Breaking Defence*, March 17, 2021.
12 DARPA, "With Squad X, Dismounted Units Partner with AI to Dominate Battlespace," *DARPA News and Events*, July 12, 2019.

intelligence and autonomy" to support its "human squadmates [with] tactical electronic and kinetic support" and "informs human decision making in complex, time-critical combat situations" for an increased battlespace and sphere of impact.[13] In other words, the technology is intended to provide the on-the-ground human squad with better and faster intelligence to act on for a range of aspects, including targeting.

A similar human-machine teaming aim underpins the U.S. Army's Advanced Targeting and Lethality Automated System (ATLAS), although with some scope for more increased kinetic autonomy. The ATLAS system's goal is to leverage advancements in Machine Learning and computer vision for integration into ground vehicles, so that acquisition, identification and engagement of targets can take place three times faster than currently possible with manual processes. For this, "[t]he ATLAS will integrate advanced sensors, processing and fire control capabilities into a weapon system to demonstrate these desired capabilities."[14] The explicit capacity to autonomously acquire, identify and engage targets through AI capabilities has raised some concerns that the U.S. is advancing their Lethal Autonomous Weapons programme despite DoD stipulations not to do so.[15] Since then, the U.S. Army has gone to great lengths to dispel the idea that ATLAS is the flagship programme for lethal autonomous tanks and other ground vehicles, rather, the narrative has shifted to highlight the human-machine teaming capabilities and stress that the machine has no lethal autonomy as such. In line with DoD Directive 3000.09, it cannot pull the trigger. As Army engineer Don Reago puts it, "[e]nvision it as a second set of eyes that's just really fast, [like] an extra soldier in the tank."[16] The programme employs machine learning algorithms and vast data sets to sift through sensory input data and to provide a recommendation for a potential target. The system itself does not determine whether the object of identification is hostile or not, it merely presents a "list of 'objects of interest' from which the human operator can choose."[17]

---

13 BAE Systems, "BAE Systems selected to provide autonomy capabilities for DARPA's Squad X Program," *BAE Systems Newsroom*, June 2, 2020.
14 Department of Defense, "Industry Day for the Advanced Targeting and Lethality Automated System (ATLAS) Program," February 2019.
15 See Kristin Houser, "US Military: Our 'Lethality Automated System' Definitely isn't a Killer Robot", *Futurism*, March 6, 2019; and Sydney Freedberg Jr, "Fear and Loathing in AI," *Breaking Defense*, March 6, 2019.
16 Don Reago, quoted in Sydney Freedberg Jr, "ATLAS: Killer Robots? No. Virtual Crewman? Yes," *Breaking Defense*, March 4, 2019.
17 Reago, "ATLAS: Killer Robots?"

The Pentagon's pioneering military AI programme is perhaps also its most well-known to date. The Algorithmic Warfare Cross-Functional Team—better known as Project Maven—has been in operation since 2017. Similar to ATLAS, Project Maven's main aim is to harness the benefits of machine learning and computer vision to accelerate targeting (and other) decisions in challenging conflict environments. Where ATLAS is intended for ground vehicles, Maven draws on drone technology for its visual and other sensory inputs to create the ability to identify and track potential targets in real-time to give the human intelligence for kinetic engagement at speed. Here too, it is stressed that the operator remains in the loop, specifically by triggering any lethal action, but, as with the ATLAS system above, it is not difficult to imagine that systems such as these are designed to help shrink the (re-)action times within which human judgment and decision-making is necessary. This, in turn, could potentially be opening the door toward greater autonomy. The technological capabilities are there.

With the AI arms race progressing at pace, other countries also refocus their attentions toward AI as a military priority. Leading the way is China, which explicitly pursues "AI-enabled systems and autonomous capabilities" which includes the identification and selection of targets.[18] Similarly, Russian weapons manufacturers, such as for example Kalashnikov, who have produced a "fully automated combat module based on neural-network technologies that enable it to identify targets and make decisions" in 2017.[19] The UK too is investing record amounts in AI technology and has shown five years of consecutive growth in AI investment. With the rising ubiquity and normalisation of AI as a functional and possibly decisive component in the lethal targeting loop, the ethical stakes could not be higher. More so in fact, as the level of sophistication in AI weapons systems is no match for the multi-varied quirks and complexities of the real world, yet. AI systems—military or otherwise—are often marred by biases, faulty or incomplete data and suffer from brittleness and vulnerabilities in its core functionalities when put to the test "in the wild." This is particularly pertinent for environments of conflict where, the fog of war is not just proverbial, but literal, where elements of uncertainty and unexpected events dominate and where fluid relations and shifting identities make it difficult to categorise and classify environments of conflict accurately.

---

18 Elsa Kania, "AI Weapons in China's Military Innovation," in *Global China: Assessing China's Growing Role in the World*, Brookings Institute Report, April 27, 2020.

19 Frank Slijper, Alice Beck, and Daan Kayser, "The State of AI: Artificial Intelligence, the Military and Increasingly Autonomous Weapons," *PAX Report Series*, May 8, 2019, 18.

An AI system might be sufficiently reliable to identify and classify a table, if the conditions are right, but how might it determine, for example, a friend from a foe in asymmetric conflicts and make appropriate targeting recommendations or other C2 tasks in which the finer nuances of human expertise are still paramount? How can an operator or commander still exercise moral judgement in an accelerated and AI enabled kill chain? To what extent is moral agency compromised?

Military AI and Mediated Human Agency

More than ever before has speed become a key factor in strategic ideas of dominance. In the U.S.'s plans for "decision dominance," speed features not just as physical speed (in hypersonics, for example) but also as "cognitive speed of an AI offering a commander options." And more than ever before, the human—operator or indeed commander—becomes integrated into a wider system of technological automation and autonomy. Crucial here is that any AI-enabled weapons system is always a *system*, an assemblage, comprised of software processes, hardware delivery platforms and, for now, the human. The human is, in this sense, always an element of the techno-logical system, embedded within cybernetic structures that operate often beyond human intelligibility, in terms of speed and computational logic. This has always been the case for operations in highly technologically advanced militaries but is set to become much more pronounced with advances in neural networks, machine learning and increased processing power.

Operating an AI-enabled weapons system is not a simple case of command and control. Where the operator is part of a system, he/she becomes reliant on information flows produced through AI data analysis. This has implications. If an AI builds a world model based on available data, it is likely to be much more successful in closed systems, where parameters can easily be grasped as data. In the context of warfare, where parameters are likely to be less fixed, more fluid and dynamic, as indicated above, the AI system may suggest a course of action based on epistemic foundations that may be biased, incomplete or otherwise not fully appropriate to the situation. In other words, the AI system frames situational awareness for the operator in specific, computational terms that may bracket important non-computational elements, or may rest on assumptions that are not always intelligible to the human element in the system. M.R. Endsley (former Chief Scientist of the U.S. Air Force) defines situational awareness as "the perception of environmental elements and events with respect to time or space, the comprehension of their meaning, and the projection of their future status."[20] Full situational awareness is the aim for commanders and an essential aspect of the C2 function. For this baseline condition, more and faster information as a mode of power is the necessary action for dominance. In plain speech, the intrinsic logic here is one of omniscience as a remedy and pathway to victory. It is therefore no great surprise, as Lucy Suchman highlights, "that in the current moment the longstanding military desire for a solution to the 'fog of war' is invested in expanded networking and data analytics."[21] Yet as the information environment produces ever-more vast amounts of data, the fog of war extends not only to actions in the physical world, but also to the increasingly complex information landscape, which, in turn, is sought to be mitigated with more data and information. Rather than offering "near total situational awareness, AI and automation will make the fog of war much worse for warfighters," as Zach Hugh notes.[22] The fog of war is likely not eliminated with technology, rather, it becomes replaced with a "fog of the system"[23] in which the technology takes on a dominant role as provider of actionable information which the human "element" of the system must take in good faith if it is to be actioned at speed.

It is this condition of systemic technological mediation and the requirement to execute a recommended action within not minutes but seconds which complicates moral agency and meaningful human control considerably in contemporary visions for war. For a moral agent to become responsible agents, certain parameters must be given. Following John Martin Fischer and Mark Ravizza, only a human can be at the heart of having "guidance control" over moral acts, and with this, an understanding that moral responsibility is a *social* practice. Moral agency means the capacity not just to have, but to *take* moral responsibility. This, in turn requires that the moral agent understands himself as such—as a moral agent within a social setting of values and expectations; that he has adequate knowledge to act as a moral agent and, importantly, that the moral decision-mechanisms

---

20 Quoted in Lucy Suchman, "Algorithmic Warfare and the Reinvention of Accuracy," *Critical Studies on Security* 8, no.2 (2020), 175-87, at 178.
21 Suchman, "Algorithmic Warfare."
22 Zach Hughes, "Fog, Friction, and Thinking Machines," *War on the Rocks*, March 11, 2020.
23 Hughes, "Fog, Friction."

is his to own.[24] In military settings, this relationship is, of course, already somewhat complicated, as the soldier and operator are part of a chain of control, command and distributed responsibility. As such, lines of responsibility are always already somewhat diffused. This diffusion is, however, further complicated with an artificial agent in the mix.

The cognitive asymmetry between humans and AI systems produces outcomes that are invariably skewed in favour of machine decisions. Cognitive psychology attests to this condition as "automation bias" in decision making, whereby "humans have a tendency to disregard or not search for contradictory information in light of a computer-generated solution that is accepted as correct and can be exacerbated in time critical domains."[25] As Cummings notes, this automation bias may be benign in situations where technological autonomy and automation is used for mundane and repetitive tasks, but when lethal decisions are at stake, automation bias risks the loss of situational awareness and may promote the degradation of important skills necessary for the ethical conduct of warfare, as human operators are unable to form an appropriate mental model in conditions of accelerated action. Even in events where it is the explicit task of the human to use and oversee technology as a tool, and intervene in a situation where the technology might fail or produce adverse outcomes, the risk is high that the human is unable to develop an appropriate mental model to help "overcome system failure."[26] The authority of knowing the parameters for the right decision falls to the technological system, when the human has not enough knowledge to appropriately assess the situation on the ground, when the human has only limited, if any, insight into the black box of algorithmic computations that underpin the recommended decision, and, when these processes are time-critical—to unfold in matters of seconds, not minutes.

As discussed above, the promise of super-human speed, efficiency and cognitive superiority is AI's greatest allure for military operations, and its core logic. Speed, as the old Sun Tzu dictum goes, "is the essence of war. Take advantage of the enemy's unpreparedness ... strike him where he has taken no precautions." With accelerated sensor to shooter timelines for which AI

---

24 John Martin Fischer and Mark Ravizza, *Responsibility and Control: A Theory of Moral Responsibility* (Cambridge: Cambridge University Press, 1998), 207-39.
25 Missy L. Cummings, "Automation Bias in Intelligent Time Critical Decision Support Systems," Conference Proceedings of the American Institute of Aeronautics and Astronautics, AIAA 2004-6313, September 20-22, 2004.
26 Richard Breton and Éloi Bossé, "The Cognitive Costs and Benefits of Automation," NATO RTO-MP-088, October 2003, 3.

holds the promise to shorten the time span from sensor data collection to weapons engagement from 20 minutes to 20 seconds, the fulfilment of speed as dominance seems within reach. This shortened timeline does, however, also mean that considerations on Rules of Engagement, collateral damage and other probabilities relevant to the targeting element in military action, all of which a responsible operator would have to take into consideration, is shortened quite substantially. The challenge then becomes weighing what is more important: the fulfilment of the aspiration of near-immediate impact without detection—as an advantage in warfare or maintaining space (and time) for intervention if something goes wrong (if the AI system miscalculates, or the enemy has decided to surrender) and for appropriate ethical deliberation on decisions with lethal outcomes. The two cannot coexist. There is ample evidence in non-military scenarios that the bet we make in the hopes to control super-fast AI systems through human action is ill-considered. An example that is often drawn on is the Flash Crash of 2010, where high-frequency trading algorithms eclipsed 600 points on the Dow Jones Index within minutes without any human interference. For a number of years, it was not fully intelligible to investigators what exactly had happened.

More pertinently perhaps, the deadly crash of a self-driving Uber car which killed pedestrian Elaine Herzberg in Arizona in 2018 illustrates the challenges implicit in automation bias and technological authority. In this fateful event, the autonomous car did not have sufficient data to recognise the woman who was pushing her bike across the street (not a pedestrian crossing) as a pedestrian. Investigations into the crash have shown that "the system design did not include a consideration for jaywalking pedestrians."[27] Moreover, the car's autonomous system was calibrated to initiate a one-second delay before braking in order to recalibrate an alternative path, or for the "safety driver" to take over. None of these processes and conditions were known to the "safety-driver" who, at the time of impact, was watching TV. The driver neither had the time, nor did she have the knowledge to form an appropriate mental model to intervene into a deadly event. It was the driver's responsibility to keep alert and keep her hands on the wheel at all times and ultimately, it was the driver who faced charges for the accident. However, it took two years before any responsibility for the accident could be clarified and blame could be assigned. Whether the driver should bear

---

[27] David Shepardson, "In review of fatal Arizona crash, U.S. agency says Uber software had flaws," *Reuters*, November 5, 2019.

sole responsibility, whether the under-developed software also has a role to play and whether anyone actually feels the weight of responsibility for an action in which an autonomous technology is substantially involved, these are questions to which the answers remain ill-defined.

Ethics as Technics and Moral Lacunas

A comparable issue arises in the context of autonomous weapons systems where the "responsibility gap" has been the topic of much and ongoing debate.[28] The attribution of moral agency and responsibility in the context of weapons systems is not straight forward. An influential 2017 study on moral agency perceptions and autonomous weapons confirms that military personnel, in particular, perceive autonomous weapons to have a "mind" and agency in terms of acting on morally relevant decisions.[29] This complicates ideas of responsibility and moral agency for lethal decision somewhat and the trends point toward front-loading responsibilities to accommodate technological capabilities that can also serve as an ethical agent. At present, there are a number of well-funded initiatives that seek to unlock the secret to making "moral machines," which would offer an in-built ethical code that ensures that the system could only be used in morally-sound ways. The Australian Defence Force, for example, was funding a multi-million (Australian) dollar project that takes this approach to heart and aims to build autonomous weapons systems where "ethics and the law are embedded in the AI and autonomous systems being used on the battlefield."[30] The methodologies associated with this agenda vary, from statistical opinion polls to more complex design proposals, but each assumes that it is possible to find a way to engineer ethics into a machine.

---

[28] For example, see Robert Sparrow, "Killer Robots", *Journal of Applied Philosophy* 24, no.1 (2007): 62-77; Andreas Matthias, "The Responsibility Gap: Ascribing Responsibility for the Actions of Learning Automata," *Ethics and Information Technology* 6, no.3 (2004), 175-83; John Sullins, "When is a Robot a Moral Agent?," *Machine Ethics* 6, no.12 (2006): 23-30; and Nehal Bhuta, Susanne Beck, and Robin Geiss, "Present futures: Concluding reflections and open questions on autonomous weapons systems" in *Autonomous Weapons Systems: Law, Ethics, Policy*, ed. Nehal Bhuta, Susanne Beck, and Robin Geiss (Cambridge: Cambridge University Press, 2016), 347-83.
[29] Ilse Verdiesen, "Agency Perception and Moral Values Related to Autonomous Weapons: An Empirical Study Using Value-Sensitive-Design Approach," Master's Thesis submitted to Delft University of Technology, The Netherlands, August 28, 2017.
[30] George Nott, "Killer robot campaign defector to 'embed ethics' in autonomous weapons," *Computer World*, March 11, 2019.

In the realm of autonomy in weapons systems, roboticist Ronald Arkin is among the most vocal proponents of systems that are both lethal and ethical. He and his colleagues introduced the idea of an ethical governor—a module tasked with constraining lethal action in autonomous weapons. Such a governor would be programmed with the appropriate rules of engagement and the parameters of the laws of war and would then be able to pre-select appropriate ethical decisions for the operator whether to allow the lethal decision to proceed or prevent it.[31] For reasons outlined above, the viability of such a constellation as a mode of control is problematic. But moreover, considering ethical decision-making in such terms, whereby a machine calculates the risks involved in a potentially lethal decision is to perhaps misunderstand the task of moral deliberation—in war or otherwise. Such thinking prioritises conceptions of war not as a social and political problem, but as an engineering task. This includes thinking about ethical dilemmas in the use of force in scientific-mathematical terms, bracketing those aspects that cannot be captured by or are not adequately represented through data. I discuss this shift toward ethics-as-technics at length elsewhere,[32] but it is worth stressing here again, that where ethics is embedded into the logic of technology, it tends to resemble a form of risk management which humans can draw on to justify morally difficult choices. It positions ethics primarily as a problem in need of a solution, and ethical deliberation takes a backseat to the technological mandate of efficiency and effectiveness. This move has a longer history and is by no means particular to the growing omnipresence of digital tools. But a growing digital dominance amplifies ideas of ethics as technics. In such ideas of ethics, the language of utility and efficiency dominates, humans are predominantly captured as data points, or numbers and more fluid categories relevant to war (such as injuries or trauma) are often bracketed. With this, moral blind spots are created and with greater technological autonomy, these blind spots are likely to become larger and exacerbate the suffering of those not embedded within a calculable field of reference or whose voices are marginalised and less visible.[33] And to rectify a wrong committed through technological recommendation or decision is likely to be an uphill struggle. Contesting technological authority requires political will and adequate resources which often are in short supply.

---

31 Ronald Arkin, Patrick Ulam, and Brittany Duncan, "An Ethical Governor for Constraining Lethal Autonomous Systems," *Georgia Institute of Technology Mobile Robot Laboratory Technical Report*, GIT-GVU-09-02 (2009), 1-8.
32 Schwarz, *Death Machines*.
33 Schwarz, "Technology and Moral Vacuums."

Neither war, nor ethics, are engineering problems. They are relational categories in which humans affect one another through action. In debating the ethics of autonomous weapons systems, ideas of warfare as technological and clean, a contest between equal state powers which can be won through technological superiority dominate. Yet contemporary warfare is far from that. Asymmetrical conflicts have dominated the landscape for decades. Such wars are messy, brutal, uncertain and highly dynamic. They involve humans who act and react, who become radicalised and resort to unexpected measures. If we take counter-insurgency doctrine seriously, we should acknowledge that winning over the trust of populations is a crucial aspect of ending spiralling violence. For this, populations, and the individuals that constitute them must be understood as more than data points. War is a social and political institution. It is relational, non-discrete and dynamic—neither a game of chess nor a game of battleship. Thinking ethically about war must consider these dimensions and ask whether the creeping techno-solutionism at work here might not make matters for populations on the ground worse, thereby prolonging the basis for conflict.

Current aspirations for AI as an accelerant for lethal decision-making places ethics in the margins of warfare. In assuming that AI can be used "as a tool" in lethal decisions we risk overlooking an irresolvable tension between the aim and character of AI on the one hand, and its intended use for (ethical) decision-making in armed conflict on the other, even with a human in a position of control. Making ethical decisions is difficult and cannot be folded into ideas of technological progress, speed, and efficiency. Ethics is context dependent and relational. Ethics asks us to make choices that often have no clear solution, that we cannot calculate, but instead require that we take responsibility. This is deeply uncomfortable to many and counter to the binary logic of computational systems, but a zone of moral discomfort is essential if we are to prevent technological violence from becoming our primary mode of addressing conflict. Digital technologies, and especially those within which the human is intricately embedded, are seldom just a tool that we employ at will. Rather, they carry a social power. They have the capacity to subtly shape our frames of reference for decision making. In so doing, they exert a powerful—and often invisible—influence over our modes of governance, our security practices, our justifications for violence, and our understanding of ethics as such. As warfare becomes increasingly systematic, through digital networks and algorithmic architectures, we must remember that these architectures might affect our thoughts and behaviour in important ways, eroding long-held humanistic values and reducing our capacity to engage in properly ethical deliberation.

Norbert Wiener, father of cybernetics, was attentive to this problem. In 1960, he wrote: "If we use, to achieve our purpose a mechanical agency with whose operation we cannot efficiently interfere once we have started it, because the action is so fast and irrevocable that we have not the data to intervene for the action is complete, then we had better be quite sure that the purpose put into the machine is the purpose which we really desire and not merely a colourful imitation of it."[34] Technology may have advanced since his days, but the concerns nonetheless remain the same. We would do well to take seriously the concerns raised at the advent of cybertechnologies, as they reveal a much more skeptical view on the limits of technology than seems customary today.

References

Arkin, R., Ulam, P. and Duncan, B. "An Ethical Governor for Constraining Lethal Autonomous Systems," *Georgia Institute of Technology Mobile Robot Laboratory Technical Report*, GIT-GVU-09-02 (2009), 1-8.

BAE Systems. "BAE Systems selected to provide autonomy capabilities for DARPA's Squad X Program." *BAE Systems Newsroom*, June 2, 2020. https://www.baesystems.com/en-us/article/providing-autonomy-capabilities-for-darpa-s-squad-x-program.

Bhuta, N., Beck, S. and Geiss, R. "Present futures: Concluding reflections and open questions on autonomous weapons systems" in *Autonomous Weapons Systems: Law, Ethics, Policy*, edited by Nehal Bhuta, Susanne Beck, and Robin Geiss, 347-83. Cambridge: Cambridge University Press, 2016.

Breton, R. and Bossé, E. "The Cognitive Costs and Benefits of Automation", NATO RTO-MP-088, October 2003.

Cole, C. "A bloody month in the drone wars: 7 separate drone strikes kill dozens of civilians across 4 war zones." *Drone Wars*, December 3, 2019. https://dronewars.net/tag/civilian-casualties/.

Cummings, M.L. "Automation Bias in Intelligent Time Critical Decision Support Systems." Conference Proceedings of the American Institute of Aeronautics and Astronautics, AIAA 2004-6313, September 20-22, 2004. https://doi.org/10.2514/6.2004-6313.

DARPA. "With Squad X, Dismounted Units Partner with AI to Dominate Battlespace." *DARPA News and Events*, July 12, 2019. https://www.darpa.mil/news-events/2019-07-12.

---

34 Wiener, "Some Moral and Technical Consequences of Automation," 87.

Department of Defense. "Industry Day for the Advanced Targeting and Lethality Automated System (ATLAS) Program." DoD Notice ID W909MY-19-R-C004, Sam.gov (beta), originally published February 11, 2019. https://beta.sam.gov/opp/6b5d5aeb584c667d4e6f5103bf6acac6/view?keywords=%22Advanced%20Targeting%20and%20Lethality%20Automated%22&sort=-relevance&index=&is_active=true&page=1.

Fischer, J. and Ravizza, M. *Responsibility and Control: A Theory of Moral Responsibility*. Cambridge: Cambridge University Press, 1998.

Freedberg Jr., S. "Army's New Aim is Decision Dominance." *Breaking Defence*, March 17, 2021. https://breakingdefense.com/2021/03/armys-new-aim-is-decision-dominance/.

Freedberg Jr, S. "ATLAS: Killer Robots? No. Virtual Crewman? Yes." *Breaking Defense*, March 4, 2019. https://breakingdefense.com/2019/03/atlas-killer-robot-no-virtual-crewman-yes/.

Freedberg Jr, S. "Fear and Loathing in AI: How the Army Triggered Fears of Killer Robots." *Breaking Defense*, March 6, 2019. https://breakingdefense.com/2019/03/fear-loathing-in-ai-how-the-army-triggered-fears-of-killer-robots/.

Houser, K. "US Military: Our 'Lethality Automated System' Definitely isn't a Killer Robot." *Futurism*, March 6, 2019. https://futurism.com/the-byte/us-military-denies-autonomous-killer-robots.

Hughes, Z. "Fog, Friction, and Thinking Machines." *War on the Rocks*, March 11, 2020. https://warontherocks.com/2020/03/fog-friction-and-thinking-machines/.

Kania, E. "AI Weapons in China's Military Innovation" in *Global China: Assessing China's Growing Role in the World*. Brookings Institute Report, April 27, 2020. https://www.brookings.edu/research/ai-weapons-in-chinas-military-innovation/.

Maggio, E. "Putin believes that whatever country has the best AI will be 'the ruler of the world.'" *Business Insider*, September 4, 2017. https://www.businessinsider.com/putin-believes-country-with-best-ai-ruler-of-the-world-2017-9?r=US&IR=T.

Matthias, A. "The Responsibility Gap: Ascribing Responsibility for the Actions of Learning Automata," *Ethics and Information Technology* 6, no.3 (2004), 175-83.

McManus, D. "Trump's brand of war is killing more civilians than before." *Los Angeles Times*, September 8, 2019. https://www.latimes.com/politics/story/2019-09-07/trumps-shameful-rules-of-engagement-are-killing-civilians.

Nott, G. "Killer robot campaign defector to 'embed ethics' in autonomous weapons." *Computer World*, March 11, 2019. https://www.computerworld.com/article/3457068/killer-robot-campaign-defector-to-embed-ethics-in-autonomous-weapons.html/.

Schwarz, E. "Prescription Drones: On the Techno-biopolitical Regimes of Contemporary 'Ethical Killing'," *Security Dialogue* 47, no. 1 (2016), 59-75.

Schwarz, E. "Technology and Moral Vacuums in Just War Theorising", *Journal of International Political Theory* 14, no. 3 (2018): 280-98.

Schwarz, E. *Death Machines: The Ethics of Violent Technologies*. Manchester: Manchester University Press, 2019.

Shanahan, J. "Lt. Gen. Jack Shanahan media briefing on A.I.-related initiatives within the Department of Defense." *US Department of Defense*, August 30, 2019. https://

www.defense.gov/Newsroom/Transcripts/Transcript/Article/1949362/lt-gen-jack-shanahan-media-briefing-on-ai-related-initiatives-within-the-depart/.

Shane, S., and Wakabayashi, D. "The business of war: Google employees protest work for the Pentagon." *New York Times*, April 4, 2018. https://www.nytimes.com/2018/04/04/technology/google-letter-ceo-pentagon-project.html.

Shepardson, D. "In review of fatal Arizona crash, U.S. agency says Uber software had flaws." *Reuters*, November 5, 2019. https://uk.reuters.com/article/uk-uber-crash/in-review-of-fatal-arizona-crash-u-s-agency-says-uber-software-had-flaws-idUKKBN1XF2HC .

Slijper, F., Beck, A. and Kayser, D. "The State of AI: Artificial Intelligence, the Military and Increasingly Autonomous Weapons." *PAX Report Series*, May 8, 2019. https://www.paxforpeace.nl/publications/all-publications/the-state-of-ai/.

Sparrow, R, "Killer Robots", *Journal of Applied Philosophy* 24, no.1 (2007): 62-77.

Strout, N. "Inside the Army's futuristic test of its battlefield artificial intelligence in the desert." *C4ISRNET*, September 25, 2020. https://www.c4isrnet.com/artificial-intelligence/2020/09/25/the-army-just-conducted-a-massive-test-of-its-battlefield-artificial-intelligence-in-the-desert/.

Suchman, L. "Algorithmic Warfare and the Reinvention of Accuracy", *Critical Studies on Security* 8, no.2 (2020): 175-87.

Sullins, J., "When is a Robot a Moral Agent?" *Machine Ethics* 6, no.12 (2006): 23-30.

United Nations. "Report of the 2019 Session of the Group of Governmental Experts on Emerging Technologies in the Area of Lethal Autonomous Weapons Systems." CCW/GGE.1/2019/3, September 25, 2019. https://undocs.org/en/CCW/GGE.1/2019/3/.

Verdiesen, I. "Agency Perception and Moral Values Related to Autonomous Weapons: An Empirical Study Using Value-Sensitive-Design Approach." Master's Thesis, submitted to Delft University of Technology, The Netherlands. August 28, 2017.

Wiener, N. *The Human Use of Human Beings*. Boston: Houghton Mifflin, 1954.

Wiener, N. "Some Moral and Technical Consequences of Automation," *Science* 131, no.3 (1960): 1355-58.

CONCLUSION

David Whetham and George Lucas

What have we learned, collectively and summatively, from our contributors concerning ethics and cyber warfare? Let us enumerate some of the specific conclusions and try to draw together compatible threads of argument.

In **Part One**, contributors help us think more carefully about what cyber conflict represents, and under which categories of governance it is most properly subsumed. In **Part Two**, contributors turn to the examination of cyber operations that appear without question to fall below the threshold of conventional war, such as sabotage, surveillance, and PSYOPs. **Part Three**, in conclusion, takes us into the realm of the speculative, attempting to discern the present and likely future impact of artificial intelligence on the cyber domain and the additional moral dilemmas it might come specifically to present.

Unifying themes run throughout the chapters, despite the different topics and foci. Most of the authors seem to agree that traditional and revisionist just war theories are collectively sufficient to focus on the moral principles that should guide (and limit) the waging of cyber conflict. From within that perspective, Michael Gross's category of sub-threshold war in the second chapter seems to capture the collective intuitions of most contributors that cyber conflict in its many forms generally fails to rise to the level of physical effects-based use of armed force in conventional war. Hence, the problems of deciding who is justifiably liable to cyber attack, when to launch an attack, and the most appropriate means and methods to inflict such attack are amply guided by the earlier reflections of Michael Walzer, Helen Frowe, and numerous other classical and revisionist just war theorists on the use of force short of full-scale war, or *jus ad vim*.

One might say in sum that, apart from a distinct conversation about the promise and perils of introducing artificial intelligence into the mix in Part Three, the essays attempt to cash out the aforementioned general insights within a range of specific situations, activities, and forms of conflict that characterize the terrain of cyber sub-threshold warfare generally. Even though their use has heretofore fallen largely within the context of ongoing kinetic hostilities, robotic weapons systems are nevertheless included within the purview of this discussion, since individual autonomous lethal weapons systems are networked and controlled by software that is itself vulnerable to cyber attack, even more so apparently when these networks are enhanced with artificial intelligence so as to move the weapons systems themselves ever farther from the realm of semi-autonomous to more fully autonomous operation beyond the limits of meaningful human control. Finally, from the opening chapter onward, the contributions seem to agree that the inherent problems of cyber conflict in all these varied forms strike the contributors to this volume as stemming from their indiscriminate and widespread targeting of civilians and civilian objects and the disproportionate harm and suffering they are capable of inflicting far beyond the boundaries or scope of the intended attack.

Let us turn then more closely to some of the findings and insights in the individual essays. Following their brief summary account of contemporary (post-revisionist) understanding of the essential just war theory criteria for enjoining (*jus ad bellum*) and fighting a war (*jus in bello*), Allhoff and Milgram proceed to apply these to questions surrounding the waging of cyber conflict. Their difficulty in doing so stems from attempting to identify acts of aggression that, like physical or kinetic attacks, would rise to the level of *casus belli*. Specifically, they encounter difficulties in assessing the equivalence and commensurability of conventional and cyber events, especially when attempting to toggle between categories of lethal and sub-lethal attacks. Lethality usually supports equivalence between conventional and cyber events, since, despite any specific proximate cause, whether cyber or conventional, death finally is death. But the cause of death may still prove relevant in other respects, as when death results indirectly from other cyber or conventional activities. The authors in this regard compare cyber crime (such as a ransomware attack on a medical facility, which happens to have a secondary effect of delayed medical treatment and death of patients) with a conventional example (dam-busting by an enemy nation that results in collateral civilian casualties). Because the latter constitutes a physical act of war, might not the former plausibly also fall under this heading?

The authors finally conclude that "of the *ad bellum* requirements, just cause and proportionality are the ones most likely to pose complex interpretive issues for cyber warfare." While, under *in bello*, the authors recognize that, with marked exceptions (like Stuxnet), cyber warfare is probably more indiscriminate overall, to the extent that it engages noncombatants more readily and more often than kinetic attacks and in a manner that is frequently out of all proportion to any genuine military necessity. These observations are fleshed out in their discussion of specific, well-known cases (for example, Estonia, Georgia, Stuxnet). Allhoff and Milgram conclude that the range of just war theory criteria in both *ad bellum* and *in bello* are difficult for cyber conflict to satisfy fully. Instead, most examples of cyber conflict fall within the "grey zone" between armed kinetic attack and espionage.

Michael Gross then proposes to subsume most cyber conflict under the rubric of *jus ad vim*, which Michael Walzer suggests should be used to evaluate uses of kinetic force short of full-scale war. Gross attempts to interpret cyber conflict through this alternative lens. Under UN Charter Article 51, he writes, nations have the right of self-defense against armed attack, "where 'armed attack' speaks to both the means and the consequences of an attack. The means are usually kinetic—bombs, tanks, missiles, and guns—while the consequences...must include territorial intrusions, human casualties or considerable destruction of property." Lacking most of these, cyber "attacks" fail to trigger the Article 51 threshold. Gross therefore proposes that we understand most forms of cyber conflict as sub-threshold warfare, in part to emphasize the high degree of political aggression and coercion involved in these cyber operations. He wonders, "Do some attackers have stronger rights than others to wage cyber warfare? And, are some attackers or targets more culpable and less innocent than others?" His key observation in answer is:

> Attribution for cyber attacks is essential to assess proportionality. When unknown agents interfere with elections, disrupt air traffic control, steal proprietary information, or destroy data, we are far afield of the law of armed conflict. Without knowing the attacker's identity, how can we know if the target responds justly or proportionately? Proportionality weighs military advantage against civilian deaths, but there are no civilian deaths in these cases. Distinction demands attention to civilian immunity, but immunity only extends to injury, loss of life, and extensive property damage. Cyber attacks may cause none of these but still be violently disruptive. In this environment, we will see that proportionality means like-kind tit-for-tat responses rather than the overwhelming response typical of armed conflict. As such, sub-threshold cyber warfare is symmetrical even when the [opposing] sides are not, and can level the playing field in a way conventional warfare cannot.

Gross ultimately concludes, with careful attention to applicable statutes of international law, that sub-threshold cyber operations are more akin to economic sanctions, as harmful but not necessarily lethal or destructive penultimate resorts to force and coercion that, in certain cases, should actually be encouraged as measures to insure that full-scale kinetic warfare remains a last resort between nations. While such measures can constitute a *casus belli*, in most instances they do not. As a result, once again, cyber conflict is more appropriately subsumed under the category of *jus ad vim* rather than *jus ad bellum*, and is subject to the same moral considerations and restraints as are economic sanctions (perhaps causing less direct harm to non-combatants). Interestingly, when considering a victim's right to employ countermeasures or some kind of deterrent retaliation, Gross takes a stronger view than the Tallinn Manual on the impact of cyber operations that disrupt email, communications, and even electoral integrity or confidence in government. These, he argues, can sometimes function as grave forms of terrorism and are aimed to achieve similar effects (even if non-violently). Hence, they may legitimately be met with stronger forms of reprisal than the Tallinn manual currently permits. Absent these, he warns, the ability of cyber aggressors to sow mistrust, fear, political divisions, and economic instability and to contribute to overall ungovernability within a target nation's borders will rapidly threaten to inflict the same kinds of widespread harm as conventional war.

Finally, Michael Skerker considers the rights of those individuals and collectivities (including nations) that are victimized in a cyber attack. Newly emergent, novel technological means (as he demonstrates with a series of precise historical examples) do not alter the moral calculus required of combatants in war time, particularly their *in bello* duties to pursue military objectives in accordance with the twin principles of discrimination and proportionality of means. These duties are grounded in the concept of each human person's natural rights, upon whose conceptual origins and foundations Skerker offers a concise and eloquent synopsis before turning to the complexities of respecting them during cyber conflict. He observes: "it is true, that as a matter of fact, cyber warfare engineers have the technical means to threaten or violate these rights with relative ease, potentially without reliable attribution. Yet just as it is a violation of [one's] rights if [one] is shot by a soldier; if [one's] local hospital is bombed; or if a foreign agent reads [one's] diary; so too are [that same person's] rights violated if a foreign actor does these sorts of things in a cyber operation, be it covert or overt." The manner in which the most basic human rights (life, property, privacy, etc.) of noncombatants and even enemy combatants may sometimes be overridden or forfeited in specific circumstances, while serving as constraints on military action in most other conceivable circumstances, Skerker concludes, is in no way dependent upon the novel technological means through which those rights may be placed in jeopardy. Thus, the same considerations (discrimination, proportionality, and exceptions regarding unintended but unavoidable collateral damage that may sometimes be recognized) remain invariant, even when contemplating a varied range of cyber attacks.

In **Part Two**, devoted to cyber operations that fall below the threshold of conventional war, Richard Schoonhoven offers a further exploration of that boundary, with attention to the case of Israel's "Operation Orchard" raid against Syria in 2007. Apart from the creative use of advance cyber attacks on the Syrian air traffic control system, he wonders, is there anything genuinely new or different from conventional war to be concerned about?

One slight difference, Schoonhoven notices, is the impact of newly available cyber tactics as were used in this case on proportionality and the so-called "threshold problem:" that with such tactics available, wars become proportionately less destructive and therefore easier to fight, thereby violating the *jus ad bellum* restraint of "last resort." But the larger if more subtle problems, he contends, are posed by *jus in bello* considerations. The targets of cyber attacks are often deliberately either dual-use or wholly civilian facilities and capabilities, like GPS, financial sectors, or an electric power grid, causing foreseeable and often unjustifiable harm to noncombatants. Cyber weapons themselves can be "wonky" or (his term) "buggy," containing flaws and vulnerabilities unknown to their designers ("zero-day" defects), or they may be capable of reverse engineering and unanticipated proliferation. Then there is the whole other dimension of AI-enabled cyber and the ability to engineer ever-greater autonomy into mechanical weapons themselves. The one bright spot falls under the category of *jus post bellum*: damage from cyber attacks is often (but not always) reversible, enabling far more rapid restoration and reconstruction following a conflict than the damage done by conventional warfare. Hence, Schoonhoven concludes, "If more and more conflict happens online, as it were, perhaps there will be less killing and physical destruction. But there is also potential for cyber to occasion widespread suffering and misery, if it substantially increases the number of wars that are fought or expands the scope of those who are caught up in those wars." Those changes, he concludes, warrant constant and careful scrutiny.

Jeremy Davis focuses on the problem of cyber-sabotage. Sabotage itself, he argues, is "an act that aims to weaken, damage, destroy, or otherwise frustrate the use of equipment, systems, environment, or other conditions necessary or useful" to the victim. Davis more questionably includes saboteur anonymity and lack of reliable attribution as essential features of sabotage. Somewhat in contrast, it seems to us that lack of clarity about the agents of sabotage is something that is occasionally (and increasingly) an element, though not an essential feature of sabotage itself. When U.S. Navy "frogmen" (SEALS) famously dynamited Japanese fuel storage facilities on enemy-occupied Pacific islands, the Japanese had little doubt who was responsible, although they were assuredly denied the use of those vital resources. Sabotage during the Cold War was usually attributed correctly by the victim to the principal adversary or a proxy. Perhaps it is closer to the intended point to remark, once again, that often, but not always, the victims of cyber sabotage are not aware that they have been sabotaged, and they instead mistakenly classify at least some of the resulting instances as random accidents. Otherwise it may be more correct to say, as Davis himself later surmises, that plausible deniability and lack of reliable attribution are essential features of effective sabotage *when the adversaries are not formally at war* or otherwise likely to enter into full-scale armed conflict. This element, plus undetectability and scale, while not unique to cyber sabotage, are especially powerful features accompanying it. Sabotage (as we learn in subsequent chapters) is a highly effective method of *preemptive or preventive attack*, seriously disrupting an adversary's essential military operations without affording a clear justification for retaliation.

In his subsequent more precise consideration of the conditions surrounding an increasing reliance on cyber sabotage, Davis essentially resurrects categories of "just cause" for use of force that were an essential feature of classical just war doctrine in the seventeenth and eighteenth centuries (for example, in the earliest works on international law by Vitoria, Gentili and Vatel), but gradually fell into disfavor. Preventive attacks are directed against a plausible future threat, but contemporary international law holds that such threats must first be actually carried out by an adversary in order to justify acts of self-defense (importantly, not retaliation or retribution) in response. In addition to this *jus ad bellum* deficiency, cyber sabotage poses twin *jus in bello* dilemmas as well: the successful act of sabotage may (1) unintentionally cause further indiscriminate collateral damage (responsibility for which must surely fall upon the unidentified saboteur), or (2) might cause the victim to engage in reprisals that are directed against the wrong parties. These *jus in bello* dilemmas place a burden of responsibility upon the saboteur for their scrupulous avoidance, while the *jus ad bellum* consideration raises the question (addressed by Walzer and others under the aforementioned new category of *jus ad vim*) of whether such preventive attacks, in the cyber case at least, might be justifiable forms of anticipatory war.

On a slightly different tack, Col. Ed Barrett takes up the more general problem of cyber operations (including, but not limited to sabotage) that fall into what policy-makers now term "war in the grey zone," or (as in Gross earlier) sub-threshold war. He offers an invaluable insider's perspective on the evolution, since 1990–91 of thinking in defense planning circles involving future preparation for the outbreak of major regional conflicts. This "six-phase" planning for what Vice Admiral Arthur K. Cebrowski initially termed "net-centric warfare" effectively managed subsequent conventional conflicts while helping prevent those conflicts from breaking out as full-scale war between the great powers. But that very success over the ensuing years "also has driven adversaries like China and Russia to engage in continuously harmful activities within a 'grey zone' between peace and war—activities against which our forces and plans are ill-suited to defend." The initial "Phase 0" operations were intended to use military theater presence and humanitarian assistance to maintain stability and avoid conventional conflict. But cyber operations by these two U.S. adversaries (including theft of intellectual property, trade, and defense secrets by China and massive campaigns of destabilizing misinformation by Russia) have created situations in which the U.S. is ill-positioned to respond in ways that would likely prove even worse, and impossible to justify.

Barrett outlines five cases of adversarial "grey zone" interference that help illustrate this impasse: (1) disabling of a municipality's computer-aided dispatch (CAD) system; (2) the placement of logic bombs within critical infrastructure; (3) cybercrime and cyber espionage; (4) voting machine tampering; (5) a real-life case of COVID vaccine disinformation. These dramatic cases raise troubling ethical questions: for example, are the avoidable deaths stemming from disabling attacks on CAD systems in major cities, or logic bombs planted in a nation's infrastructure, or crippling economic harm inflicted on the victim state's population worthy of lethal kinetic response? If not, what kind of response, if any, ought to be undertaken? It may come as a surprise that the debate in defense circles is precisely over such questions as the primary challenges to be faced in crafting appropriate countermeasures to these kinds of cyber "grey zone" attacks. Barrett's own subsequent discussion of answers to the ethical challenges posed is itself well-grounded, judicious, and highly reassuring regarding the sophistication of informed thought taking place within the security forces of the U.S. and its allies.

Peter Lee assesses the impact of cyber surveillance and intrusion on the notion of state sovereignty and territorial integrity. The latter doesn't amount to much within a domain that doesn't recognize or respect geographical boundaries. The overriding question is whether or not aggressive cyber surveillance and data extrusion, among other things, constitute grounds for preventive war (kinetic retaliatory response). Heretofore, in other pre-cyber or non-cyber situations involving such forms of espionage, the answer has been pretty clearly "no!" But military cyber surveillance, Lee argues, "appears to have two direct links to just war ethical assessment: direct and indirect. Direct relevance exists where the surveillance breaches the territory and sovereignty of another state in what may or may not be an act of war or application of force-short-of-war. Indirect relevance emerges when cyber surveillance provides the means by which assessments about a prospective enemy's actions and intentions can be made." The mixed motives, combined with scale of damage and harm inflicted through cyber means, may have inaugurated a new regime of *jus ad vim*, in which justifiable and proportionate response by a victim of aggressive cyber espionage may come to include kinetic responses that would previously have been considered unthinkable. And this returns us to Barrett's earlier observation regarding the present inadequacy of U.S. and allied policy toward these intrusions: we are compelled either to refrain from meaningful countermeasures or to undertake responses that themselves would otherwise seem morally unjustifiable.

Another traditional form of information "warfare" that has taken on a new dimension of significance in the cyber era is psychological operations, or PSYOP. Aggressive cyber-enabled PSYOP by Russian, Chinese, and possibly Iranian agents since 2016 have turned a millennia-old practice of sub-threshold conflict into a seemingly serious new game-changing consideration for its victims in the U.S. (multiple election interference), in the UK (where the Brexit vote may have been influenced), and in other allied democratic nations in which key internal electoral decisions or political policies may have been influenced or altered in a manner gravely affecting the resulting global balance of power. Is there, Adam Henschke asks, a difference in kind here that carries any novel moral significance?

The answer lies both in the cyber-enabled means employed (CE-PSYOP, such as social media), and in the kinds of proximate and ultimate political goals such means may now enable. It is not merely, for example, that one political candidate's character was besmirched, or even that a single election outcome might have been transformed. The ultimate loss of faith in, for example, the policies and procedures democratic governments (so as to undermine their continued ability to govern) may constitute an ultimate goal of such operations that would not have been realistically attainable without the cyber-enabled means. This in turn suggests that aggressive CE-PSYOP may no longer constitute a prelude or companion to kinetic warfare, but a viable replacement for it. Henschke argues therefore that it would become subject to traditional *jus ad bellum* criteria—criteria with which the most prominent and prevalent recent forms of PSYOP fail to comply.

Even so, might not this sort of conflict (as *jus ad vim*) be preferrable to full-scale kinetic warfare, on the grounds of both proportionality and last resort? Perhaps, but the prospect of widespread collateral damage both to unintended civilian victims and to public trust in institutions generally, he concludes, present a risk ultimately of indiscriminate and disproportionate harm that is difficult to calculate prospectively and that should therefore caution against the easy resort to such means of conflict. This, Henschke demonstrates, becomes especially complex and tangled in the case of defensive CE-PSYOP, especially when undertaken by a government against its own citizens in an effort to shield them from an adversary's PSYOP.

Part Three, in conclusion, takes us somewhat into the realm of the speculative, attempting to discern the present and likely future impact of artificial intelligence on the cyber domain and the additional moral dilemmas it might come specifically to present.

Andrew Tidmarsh of the UK's Royal Air Force offers the prospect of both AI-enhanced cyber security defenses and LAWS as potential "guardian angels" of a nation's sovereignty when the AI-enhancements for either or both are designed with a capacity to conduct and constrain their actions by means of machine-based moral reasoning. In support of his thesis, it bears mention that some underwater weapons systems, in fact, already are in use with a very rudimentary, legalistic capacity for moral constraint.[1] But indeed,

---

[1] Brutzman, Donald, Davis, Duane, Lucas, George, McGhee, Robert, "Runtime Ethics Checking for Autonomous Unmanned Vehicles: Developing a Practical Approach," *Proceedings of the 18th International Symposium on Unmanned Untethered Submersible Technology (UUST)*, Portsmouth New Hampshire, August 2013: https://savage.nps.edu/AuvWorkbench/documentation/papers/UUST2013PracticalRuntimeAUVEthics.pdf; Robert Sparrow and George Lucas, "When Robots Rule the Waves," *The Naval War College Review* 69, no. 4 (Autumn 2016): 49-78.

machine morality might prove essential, if Tidmarsh's concerns about the unreliability of human defenders is warranted. He launches a foray into recent behavioral science research focused on socially divisive hate speech, as well as on the manifestation of defective or latent-criminal responses in human agents when they are placed in trusted security situations. This lengthy examination offers evidence supporting a claim made by roboticist Ronald Arkin a decade ago (and discussed at greater length in the chapter immediately following): namely, that not only is it feasible to incorporate moral reasoning into autonomous lethal weapons design, it is important that we quickly act to do so in order to guarantee a greater degree of legally compliant behavior during armed conflict. Robots, Arkin provocatively claims, are inherently more reliable than are humans under duress to "do the right thing" (or, at very least, to comply with international law).

The challenges of designing and operationalizing moral reasoning within artificial intelligence are, of course, manifold. Tidmarsh accordingly examines at length some of the more promising lines of research, arguing that if this formidable problem can be solved, the resulting AI systems "could be employed wherever there is a machine in the loop of human decision making, whether at the strategic level of information operations, or at the tactical level of physical warfighting" where, in addition to their overall competence and effectiveness in carrying out these tasks, moral machines might prove to be more upstanding and incorruptible than human agents.

This theme is further examined in the next chapter, in which Scott Robbins assesses the prospects for designing or incorporating a capacity for human moral reasoning in AI-enhanced systems, and whether doing so might resolve the problem of "meaningful human control" in otherwise-autonomous, lethally armed weapons systems. In essence, rather than actually having humans in control over the ethical decisions confronted by lethally-armed weapons systems, the requirement of meaningful control and accountability could be met by endowing the weapons systems themselves, as an essential component of their autonomy, with the same capacities for ethical reasoning possessed by humans. Unfortunately for Tidmarsh's optimism in the prior chapter, Robbins (who has himself previously made important contributions to this ongoing debate over what we might term "autonomy and meaningful machine morality") ultimately concludes that the prospects are not very good, in part because the debate itself regarding what would count as moral reasoning embodies a fundamental category mistake. Designers confuse the ability to undertake complex moral reasoning in highly ambiguous cases with issues like "safety," risk-management, or rule-governed legal compliance. And while important, these rudimentary capacities are far from identical to, let alone do they adequately encompass, the category of "moral reasoning." The goal of meaningful machine morality is further compromised by a distinct set of flaws in AI itself, which pose their own moral challenges. These include biases both in character recognition and in judgments based upon incomplete or slanted data sets, the "black box" problem ("algorithmic opacity"), the inability to account for how the "intelligent" system reaches specific conclusions (that might be seriously in error), and similar, much-discussed limitations in AI-based reasoning that undermine its overall reliability. The best way to avoid encountering these limitations, Robbins concludes, is to understand them and refrain from deploying AI-systems in situations in which they would likely come into play—including allowing machines to make moral decisions about the use of lethal force.

So where does all this leave us in the race to incorporate ever-more applications of AI within military cyber systems and autonomous weapons? Elke Schwarz concludes this discussion by observing how the promise of AI itself has recently supplanted the wider discussion of automated weapons systems as promising a "new era" of risk-averse and ever-more accident and casualty-free warfare obtained through greater precision in targeting, discrimination of noncombatants, and reduced damage. These were the same promises, she observes, associated with unmanned systems a decade ago.

It therefore seems appropriate to inquire how this is all working out for us. Schwarz describes several U.S. Army systems currently in development, including MITS (mission intelligence tactical systems) and ATLAS (advanced targeting and lethality automated system) both of which aim to shorten the time and vastly improve the accuracy of acquiring, identifying, and engaging enemy positions. Because these systems involve machine learning and presumably could enable autonomous targeting and firing, they appear to conflict with DoD Directive 3000.09 (2012) aimed at prohibiting machines themselves to "pull the trigger" on presumed enemy targets. The Army counters that such systems in the ground combat force mix merely enable human combatants to operate more quickly and effectively. A similar case is made for "Project Maven" (the algorithmic warfare cross-functional team), aimed at improving the capabilities of aerial drones.

All these AI-enhanced weapons projects, however, currently run afoul of the same defects that mar AI systems generally: such systems—military or otherwise, she observes—"are often marred by biases, faulty or incomplete data and suffer from brittleness and vulnerabilities in core functionalities when put to the test." As Robbins also noted in the previous chapter, when such systems "get things right," they can perform quite well; but if they go awry, they go spectacularly wrong. Their common vulnerability is what military personnel call "situational awareness," the need to frame problems and model relevant data to assess overall combat environment and reasonably predict what is transpiring or about to occur. Machine learning (ML) historically keeps tripping up over this fundamental problem, harking back to arguments in the 1970s and 1980s between the late philosopher Hubert Dreyfus and renowned AI-advocate Marvin Minsky at MIT, concerning what machines can and cannot do. They inherently lack a background nexus of concerns and experiences, a "cultural horizon" that drives human intuitions and often informs what we vaguely term "common sense." Limited as they currently are to calculative rationality, no matter how powerful that capacity, or vast, accurate, and even impartial the data sets that inform their reasoning, machine algorithms lack any organic connection to their surrounding environment: they "don't care," and they lack guiding intuitions. This makes the performance of AI-enhanced systems highly unstable and unreliable in unfamiliar circumstances (such as in the midst of Clausewitz's famous "fog of war"). This same fundamental problem haunts the attempts to design "ethical governors" for robotic systems (Arkin), or otherwise incorporate "ethical reasoning" into ML: all reduce moral reasoning to calculative rationality, and they tend to treat moral dilemmas as a kind of engineering problem (for example, "optimization" of outcomes, or risk analysis).

But this is both incorrect conceptually, and wholly inadequate to guide practical performance. Schwarz's own summary of the problem lying at the heart of this volume and its contributors collectively, stands as eloquent testimony to their fundamental agreement on the promise and perils of digital technology in "the grey zone." She wisely concludes, as perhaps we should as well:

> Neither war, nor ethics, are [exclusively] engineering problems. They are relational categories in which humans affect one another through action. In debating the ethics of autonomous weapons systems, ideas of warfare as technological and clean, a contest between equal state powers which can be won through technological superiority dominate. Yet contemporary warfare is far from that. Asymmetrical conflicts have

dominated the landscape for decades. Such wars are messy, brutal, uncertain and highly dynamic. They involve humans who act and react, who become radicalised and resort to unexpected measures. If we take counter-insurgency doctrine seriously, we should acknowledge that winning over the trust of populations is a crucial aspect of ending spiralling violence. For this, populations, and the individuals that constitute them must be understood as more than data points. War is a social and political institution. It is relational, non-discrete and dynamic—neither a game of chess nor a game of battleship. Thinking ethically about war must consider these dimensions and ask whether the creeping techno-solutionism at work here might not make matters for populations on the ground worse, thereby prolonging the basis for conflict.

References

Brutzman, D., D. Davis, G. Lucas and R. McGhee. "Runtime Ethics Checking for Autonomous Unmanned Vehicles: Developing a Practical Approach," *Proceedings of the 18th International Symposium on Unmanned Untethered Submersible Technology (UUST)*, Portsmouth New Hampshire, August 2013: https://savage.nps.edu/AuvWorkbench/documentation/papers/UUST2013PracticalRuntimeAUVEthics.pdf.

Sparrow, R. and G. Lucas, "When Robots Rule the Waves," *The Naval War College Review* 69, no. 4 (Autumn 2016): 49-78.

INDEX

Algorithm 156, 161, 172–185
Artificial Intelligence 152, 161–167, 172–185, 190–201, 205–206, 209
ATLAS 193–194, 215
Autonomous weapons systems 28, 30, 69–70, 79, 103, 113, 119, 122, 156, 172, 178–181, 189–194, 198–201, 206, 214–216
Biden, Joseph 116, 121
BREXIT 131, 156–158, 212
CactusPete 111–112
Casus belli 4, 117, 119, 138, 206
Clausewitz, Carl von 164, 216
Conventional military operations 2–3, 29, 45, 53, 59–73, 75, 94, 207
China 4, 17, 19–20, 69, 95–100, 102, 110–117, 123–125, 131, 159, 190–191, 211, 212
Civilians 32–41, 52–54, 68–69, 84, 105, 135–137, 141–142, 153, 178–179, 190
COVID-19 92, 95, 99, 158–159, 211
Cyber Command (U.S.) 2, 68, 96, 108
Denial 1, 14, 21, 97, 102
Double effect 49–52, 54–55, 101
Dual-use 32–35, 39, 45, 53, 55, 67, 209

Elections 29, 41, 94, 98, 116, 121, 157, 207
Espionage 19–20, 24, 38, 60, 92, 97, 110–111, 119, 124, 207
Force-short-of-war 4, 20, 27, 34, 36, 86–87, 101, 112, 118–119, 121–122, 139, 205
Grey zone 92–94, 107–108, 207
Humanitarian Law 28–33, 59–60, 178
Iran 2, 17, 19, 21–23, 32, 35, 63–65, 70, 78–79, 81–82, 85, 131, 135, 212
Iraq 35, 67, 94, 139
ISIS 52–53
Israel 1–2, 17, 22, 27, 30, 32, 35, 60–63, 79–81, 84–85, 87, 97, 209
Jus ad Bellum 10, 12–15, 18–20, 23–24, 61–63, 83–84, 86, 92, 100, 112, 122, 130, 137–138, 141, 206, 208–209
Jus ad Vim 28–42, 86–87, 92, 101, 112, 121–123, 139, 205, 207–208, 211
Jus in Bello 9–10, 12, 16, 18, 24, 45, 53, 61, 63, 83–84, 86, 100, 112, 129, 141, 206
Just War Theory 9–24, 83–84, 118–120, 181, 206–207

Kinetic military action 4–5, 13–19, 21, 23–24, 28–29, 41, 45, 49, 51, 53, 59, 61–72, 85, 89, 94, 101, 104, 108, 129, 137, 139–142, 193–194, 206–208, 211–213

Law 15–17, 27–33, 31, 36–40, 45, 47, 52, 59–60, 75, 105–106, 112–122, 142, 144, 178, 180, 199–200, 207–208, 210, 214

9/11 4, 113

Machine Learning 3, 5, 156, 164, 173–175, 179–182, 189, 192–195, 215–216

Operation Orchard 1, 61, 67, 78–81, 87, 209

Phase Zero 92–109

Propaganda 52, 131, 133, 136, 139, 140, 143–144, 176

Proportionality 11–13, 16, 18–21, 23, 29, 33–35, 37–40, 53, 55, 62–64, 66, 69, 83–84, 100–101, 121, 137, 140–142, 181, 206–208

PSYOP 130–148

Robots 70, 178–184, 191, 214

Russia 1–2,4, 17, 20–21, 67, 94–99, 102–103, 110, 114, 116, 121, 123, 130–131, 156–157, 190, 194, 212

Sabotage 60–61, 74–91, 110

Sanctions 11, 17, 31–32, 34–35, 39, 86, 208

Sovereignty 84, 103, 112–117, 120, 212–213

Stuxnet 2, 22, 32, 61, 63–67, 78–88, 135–136, 206–207

Sub-threshold 4, 27–43, 45, 61, 65

Superintelligence 183–184

Surveillance 29, 53, 110, 128, 173

Syria 1, 27, 59, 79–80, 87, 114, 178, 209

Tallinn Manual 30, 36–37, 41, 115, 119, 142, 208

Terrorism 4, 28, 30–31, 41, 90, 135, 143, 158, 173, 176–177, 190, 208

Trump, Donald 103, 116, 130, 135, 157–159

Walzer, Michael 9–10, 27, 63, 118–119, 205, 207

Zero-days 65, 78, 81, 209

www.ingramcontent.com/pod-product-compliance
Ingram Content Group UK Ltd.
Pitfield, Milton Keynes, MK11 3LW, UK
UKHW021319180426
11947UKWH00015B/1315